KiCad - Getting Started in KiCad

A catalogue record for this book is available from the Hong Kong Public Libraries.

Published in Hong Kong by Samurai Media Limited.

Email: info@samuraimedia.org

ISBN 978-988-8381-86-9

Background Cover Image by https://www.flickr.com/people/webtreatsetc/

Contents

Chapter 1

Introduction to KiCad

KiCad is an open-source software tool for the creation of electronic schematic diagrams and PCB artwork. Beneath its singular surface, KiCad incorporates an elegant ensemble of the following stand-alone software tools:

Program name	Description	File extension
KiCad	Project manager	`*.pro`
Eeschema	Schematic editor (both schematic and component)	`*.sch, *.lib, *.net`
CvPcb	Footprint selector	`*.net`
Pcbnew	Circuit board board editor	`*.kicad_pcb`
GerbView	Gerber viewer	All the usual gerbers
Bitmap2Component	Convert bitmap images to components or footprints	`*.lib, *.kicad_mod, *.kicad_wks`
PCB Calculator	Calculator for components, track width, electrical spacing, color codes, and more…	None
Pl Editor	Page layout editor	`*.kicad_wks`

Note

The file extension list is not complete and only contains a subset of the files that KiCad works with that is useful for the basic understanding of which files are used for each KiCad unique application.

KiCad can be considered mature enough to be used for the successful development and maintenance of complex electronic boards.

KiCad does not present any board-size limitation and it can easily handle up to 32 copper layers, up to 14 technical layers and up to 4 auxiliary layers. KiCad can create all the files necessary for building printed boards, Gerber files for photo-plotters, drilling files, component location files and a lot more.

Being open source (GPL licensed), KiCad represents the ideal tool for projects oriented towards the creation of electronic hardware with an open-source flavour.

On the Internet, the home of KiCad is:

http://www.kicad-pcb.org/

1.1 Download and install KiCad

KiCad runs on GNU/Linux, Apple OS X and Windows. You can find the most up to date instructions and download links at:

http://www.kicad-pcb.org/download/

Important

The current KiCad stable builds have not been updated since 2013. As such they are missing many new features that have been added to KiCad. It is suggested that you install the unstable nightly builds which occasionally introduce new bugs, but are updated frequently and generally stable.

1.2 Under GNU/Linux

Stable builds Stable releases of KiCad can be found in most distibution′ s package managers as kicad and kicad-doc.

Unstable (nightly development) builds Unstable builds are built from the most recent source code. They can sometimes have bugs that cause file corruption, generate bad gerbers, etc, but are generally stable and have the latest features.

Under Ubuntu, the easiest way to install an unstable nightly build of KiCad is via *PPA* and *Aptitude*. Type the following into your Terminal:

```
sudo add-apt-repository ppa:js-reynaud/ppa-kicad
sudo aptitude update && sudo aptitude safe-upgrade
sudo aptitude install kicad kicad-doc-en
```

Under Fedora the easiest way to install an unstable nightly build is via *copr*. To install KiCad via copr type the following in to copr:

```
sudo dnf copr enable mangelajo/kicad
sudo dnf install kicad
```

Alternatively, you can download and install a pre-compiled version of KiCad, or directly download the source code, compile and install KiCad.

1.3 Under Apple OS X

Stable builds There are currently no stable builds of KiCad for OS X.

Unstable (nightly development) builds Unstable builds are built from the most recent source code. They can sometimes have bugs that cause file corruption, generate bad gerbers, etc, but are generally stable and have the latest features.

Unstable nightly development builds can be found at: http://downloads.kicad-pcb.org/osx/

1.4 Under Windows

Stable builds Stable builds of KiCad can be found at: http://downloads.kicad-pcb.org/archive/

Unstable (nightly development) builds Unstable builds are built from the most recent source code. They can sometimes have bugs that cause file corruption, generate bad gerbers, etc, but are generally stable and have the latest features.

For Windows you can find nightly development builds at: http://downloads.kicad-pcb.org/windows/

1.5 Support

If you have ideas, remarks or questions, or if you just need help:

- Visit the Forum
- Join the #kicad IRC channel on Freenode
- Watch Tutorials

Chapter 2

KiCad Workflow

Despite its similarities with other PCB software tools, KiCad is characterised by an interesting work-flow in which schematic components and footprints are actually two separate entities. This is often the subject of discussion on Internet forums.

2.1 KiCad Workflow overview

The KiCad work-flow is comprised of two main tasks: making the schematic and laying out the board. Both a component library and a footprint library are necessary for these two tasks. KiCad has plenty of both. Just in case that is not enough, KiCad also has the tools necessary to make new ones.

In the picture below, you see a flowchart representing the KiCad work-flow. The picture explains which steps you need to take, in which order. When applicable, the icon is added as well for convenience.

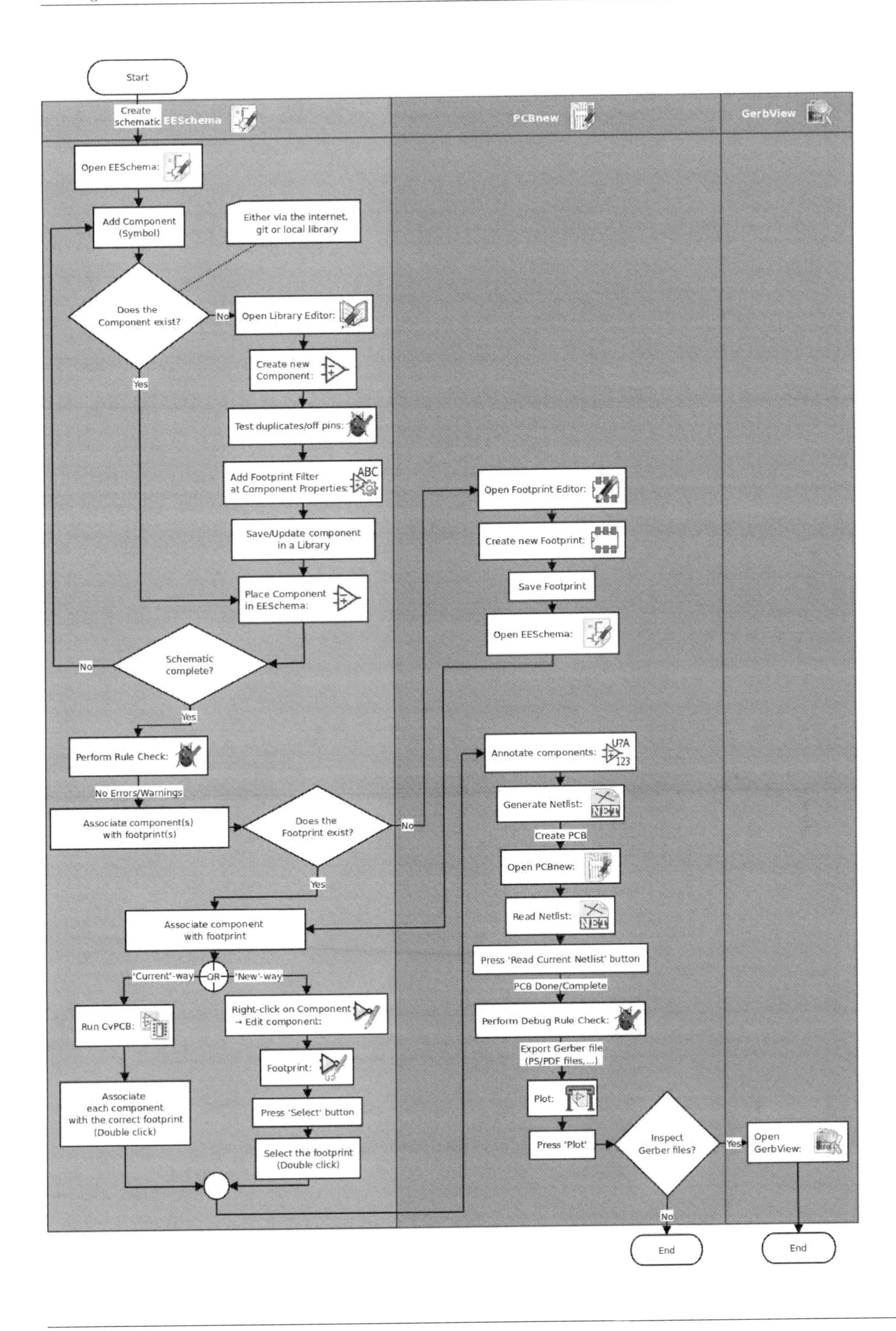
Start
Create schematic
EESchema
PCBnew
GerbView
Open EESchema:
Add Component (Symbol)
Either via the internet, git or local library
Does the Component exist?
No
Open Library Editor:
Create new Component:
Yes
Test duplicates/off pins:
Add Footprint Filter at Component Properties:
Save/Update component in a Library
Place Component in EESchema:
Schematic complete?
No
Yes
Perform Rule Check:
No Errors/Warnings
Associate component(s) with footprint(s)
Does the Footprint exist?
No
Yes
Associate component with footprint
'Current'-way
OR
'New'-way
Run CvPCB:
Associate each component with the correct footprint (Double click)
Right-click on Component → Edit component:
Footprint:
Press 'Select' button
Select the footprint (Double click)
Open Footprint Editor:
Create new Footprint:
Save Footprint
Open EESchema:
Annotate components:
Generate Netlist:
Create PCB
Open PCBnew:
Read Netlist:
Press 'Read Current Netlist' button
PCB Done/Complete
Perform Debug Rule Check:
Export Gerber file (PS/PDF files, ...)
Plot:
Press 'Plot'
Inspect Gerber files?
Yes
No
Open GerbView:
End
End

For more information about creating a component, see the section of this document titled Make schematic components in KiCad. And for more information about how to create a new footprint, see the section of this document titled Make component footprints.

On the following site:

http://kicad.rohrbacher.net/quicklib.php

You will find an example of use of a tool that allows you to quickly create KiCad library components. For more information about quicklib, refer to the section of this document titled Make Schematic Components With quicklib.

2.2 Forward and backward annotation

Once an electronic schematic has been fully drawn, the next step is to transfer it to a PCB following the KiCad work-flow. Once the board layout process has been partially or completely done, additional components or nets might need to be added, parts moved around and much more. This can be done in two ways: Backward Annotation and Forward Annotation.

Backward Annotation is the process of sending a PCB layout change back to its corresponding schematic. Some do not consider this particular feature especially useful.

Forward Annotation is the process of sending schematic changes to a corresponding PCB layout. This is a fundamental feature because you do not really want to re-do the layout of the whole PCB every time you make a modification to your schematic. Forward Annotation is discussed in the section titled Forward Annotation.

Chapter 3

Draw electronic schematics

In this section we are going to learn how to draw an electronic schematic using KiCad.

3.1 Using Eeschema

1. Under Windows run kicad.exe. Under Linux type *kicad* in your Terminal. You are now in the main window of the KiCad project manager. From here you have access to eight stand-alone software tools: *Eeschema*, *Schematic Library Editor*, *Pcbnew*, *PCB Footprint Editor*, *GerbView*, *Bitmap2Component*, *PCB Calculator* and *Pl Editor*. Refer to the work-flow chart to give you an idea how the main tools are used.

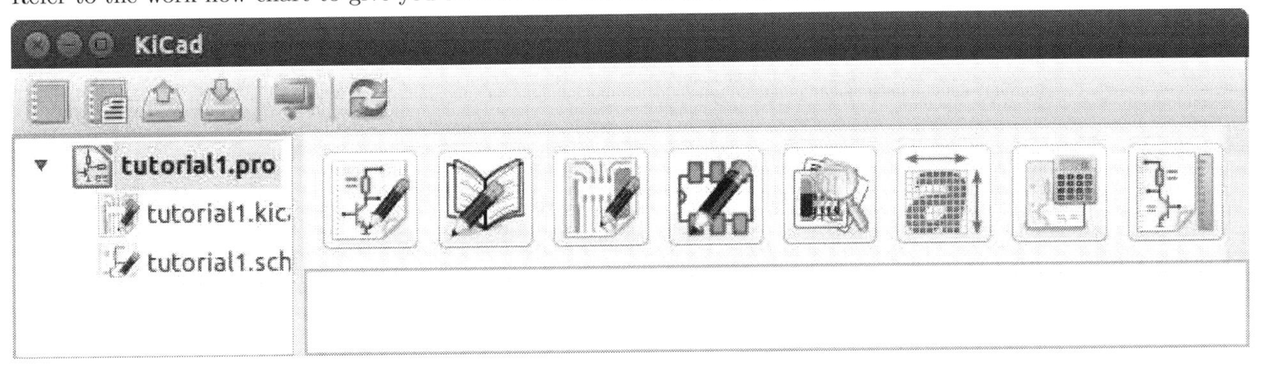

2. Create a new project: **File** → **New Project** → **New Project**. Name the project file *tutorial1*. The project file will automatically take the extension ".pro". KiCad prompts to create a dedicated directory, click "Yes" to confirm. All your project files will be saved here.

3. Let' s begin by creating a schematic. Start the schematic editor *Eeschema*, . It is the first button from the left.

4. Click on the *Page Settings* icon on the top toolbar. Set the Page Size as *A4* and enter the Title as *Tutorial 1*. You will see that more information can be entered here if necessary. Click OK. This information will populate the schematic sheet at the bottom right corner. Use the mouse wheel to zoom in. Save the whole schematic project: **File** → **Save Schematic Project**

5. We will now place our first component. Click on the *Place component* icon in the right toolbar. The same functionality is achieved by pressing the *Place component* shortcut key (*a*).

Note

You can see a list of all available shortcut keys by pressing the *?* key.

6. Click on the middle of your schematic sheet. A *Choose Component* window will appear on the screen. We' re going to place a resistor. Search / filter on the *R* of **R**esistor. You may notice the *device* heading above the Resistor. This *device* heading is the name of the library where the component is located, which is quite a generic and useful library.

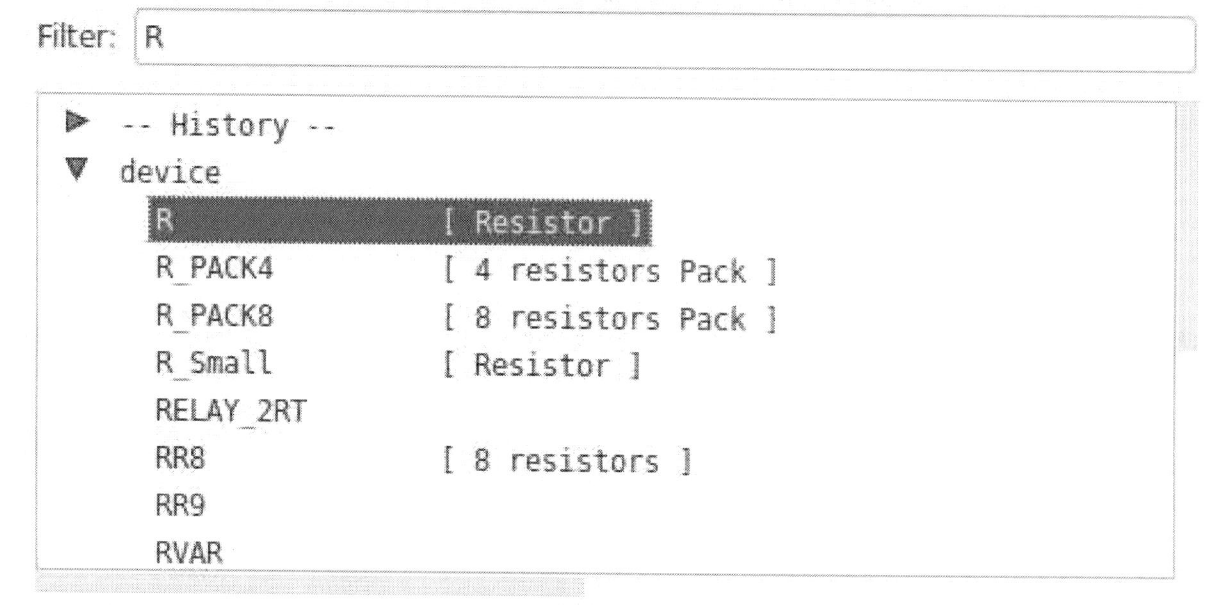

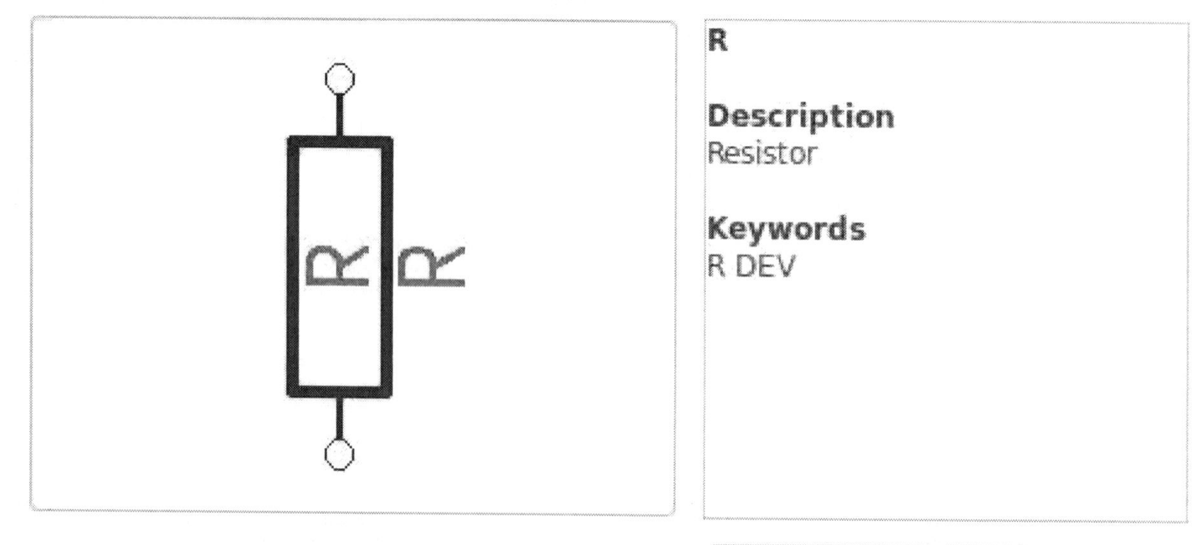

7. Double click on it. This will close the *Choose Component* window. Place the component in the schematic sheet by clicking where you want it to be.
8. Click on the magnifier icon to zoom in on the component. Alternatively, use the mouse wheel to zoom in and zoom out. Press the wheel (central) mouse button to pan horizontally and vertically.
9. Hover the mouse over the component *R* and press the r key. Notice how the component rotates.

Note
You do not need to actually click on the component to rotate it.

10. Right click in the middle of the component and select **Edit Component** → **Value**. You can achieve the same result by hovering over the component and pressing the v key. Alternatively, the e key will take you to the more general Edit window. Notice how the right-click menu below shows shortcut keys for all available actions.

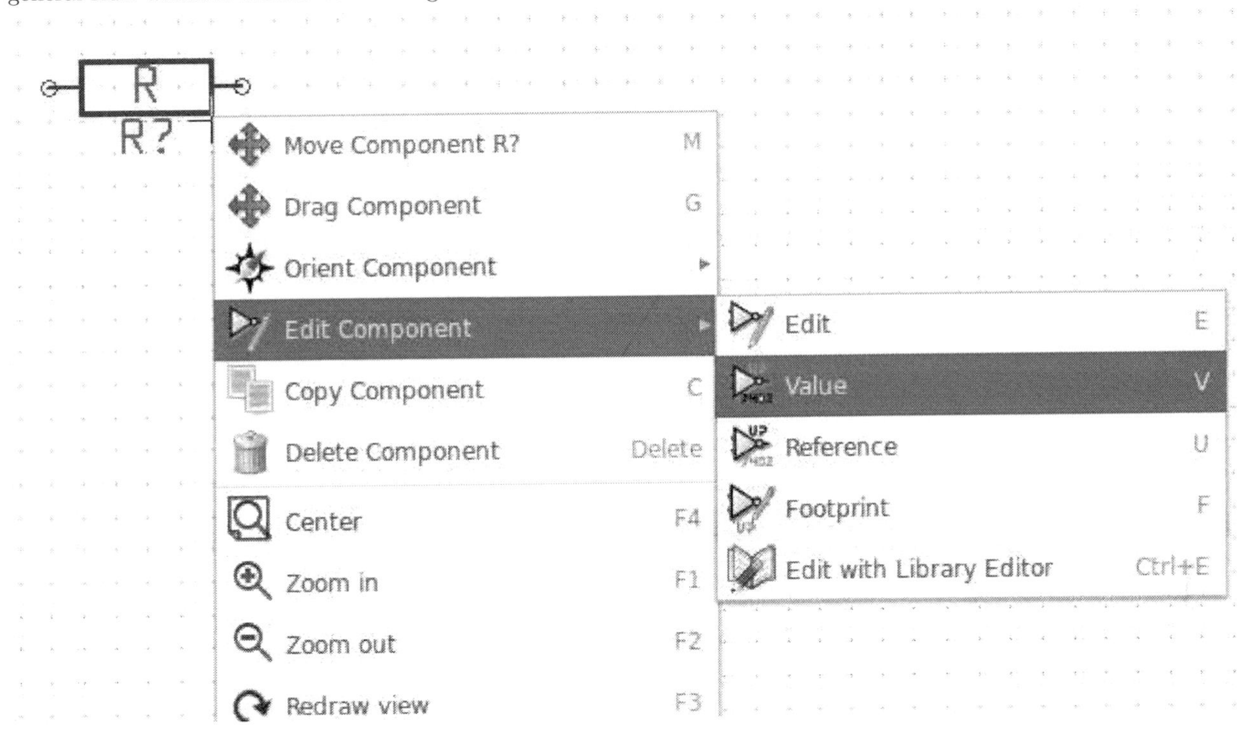

11. The Component value window will appear. Replace the current value *R* with *1k*. Click OK.

Note
Do not change the Reference field (R?), this will be done automatically later on. The value inside the resistor should now be *1k*.

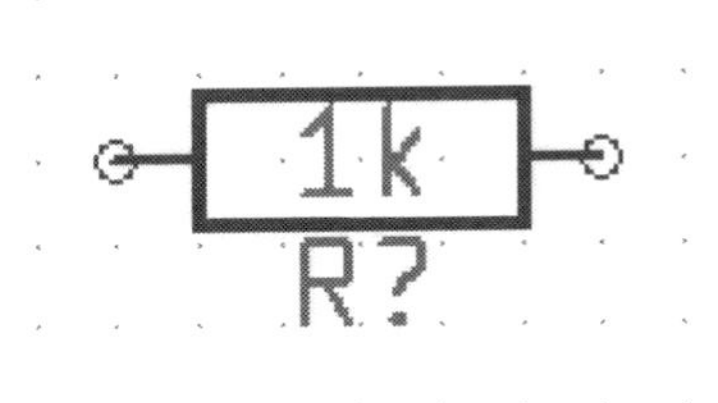

12. To place another resistor, simply click where you want the resistor to appear. The Component Selection window will appear again.

13. The resistor you previously chose is now in your history list, appearing as *R*. Click OK and place the component.

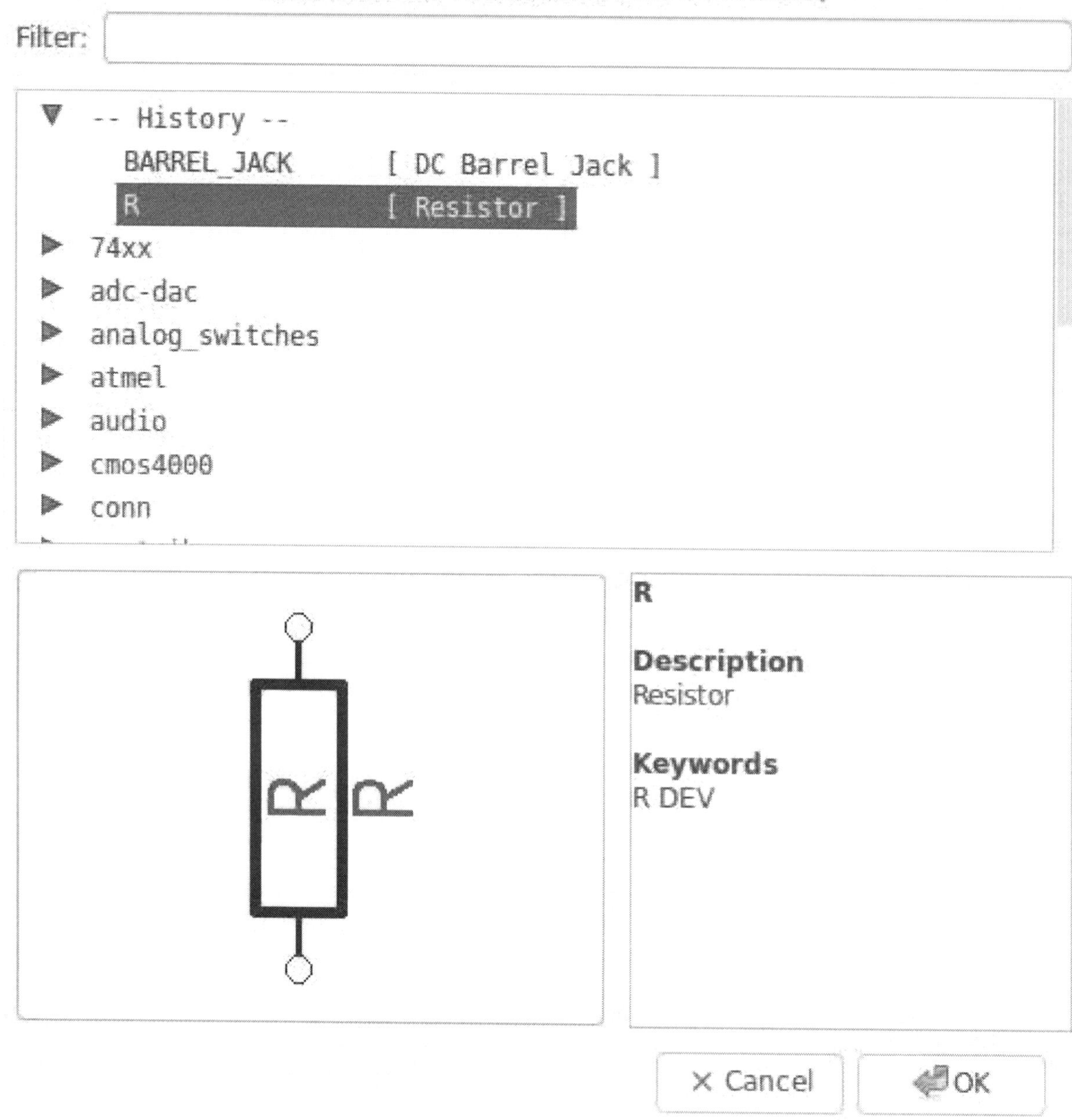

14. In case you make a mistake and want to delete a component, right click on the component and click *Delete Component*. This will remove the component from the schematic. Alternatively, you can hover over the component you want to delete and press the del key.

Note

You can edit any default shortcut key by going to **Preferences** → **Hotkeys** → **Edit hotkeys**. Any modification will be saved immediately.

15. You can also duplicate a component already on your schematic sheet by hovering over it and pressing the c key. Click where you want to place the new duplicated component.

16. Right click on the second resistor. Select *Drag Component*. Reposition the component and left click to drop. The same functionality can be achieved by hovering over the component and by pressing the g key. Use the r key to rotate the component. The x key and the y key will flip the component.

Note

Right-Click → **Move component** (equivalent to the m key option) is also a valuable option for moving anything around, but it is better to use this only for component labels and components yet to be connected. We will see later on why this is the case.

---

17. Edit the second resistor by hovering over it and pressing the v key. Replace *R* with *100*. You can undo any of your editing actions with the ctrl+z key.

18. Change the grid size. You have probably noticed that on the schematic sheet all components are snapped onto a large pitch grid. You can easily change the size of the grid by **Right-Click** → **Grid select**. *In general, it is recommendable to use a grid of 50.0 mils for the schematic sheet.*

19. Repeat the add-component steps, however this time select the *microchip_pic12mcu* library instead of the *device* library and pick the *PIC12C508A-I/SN* component instead of the *R* component from it. Before add-component, add *microchip_pic12mcu* to your Component library files by **Preferences** → **Component Libraries** and press Add button.

20. Hover the mouse over the microcontroller component. Press the y key or the x key on the keyboard. Notice how the component is flipped over its x axis or its y axis. Press the key again to return it to its original orientation.

21. Repeat the add-component steps, this time choosing the *device* library and picking the *LED* component from it.

22. Organise all components on your schematic sheet as shown below.

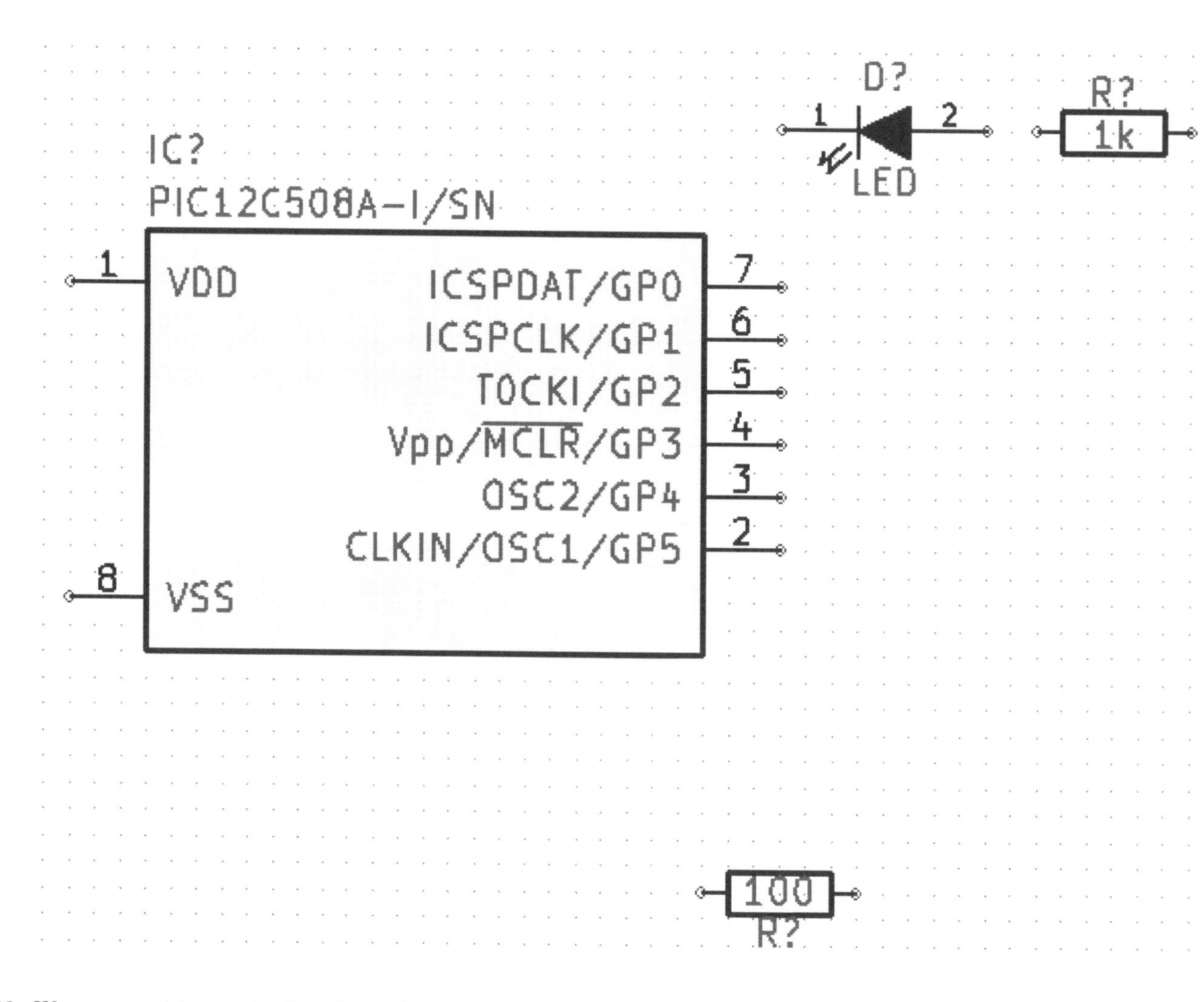

23. We now need to create the schematic component *MYCONN3* for our 3-pin connector. You can jump to the section titled Make Schematic Components in KiCad to learn how to make this component from scratch and then return to this section to continue with the board.

24. You can now place the freshly made component. Press the *a* key and pick the *MYCONN3* component in the *myLib* library.

25. The component identifier *J?* will appear under the *MYCONN3* label. If you want to change its position, right click on *J?* and click on *Move Field* (equivalent to the m key option). It might be helpful to zoom in before/ while doing this. Reposition *J?* under the component as shown below. Labels can be moved around as many times as you please.

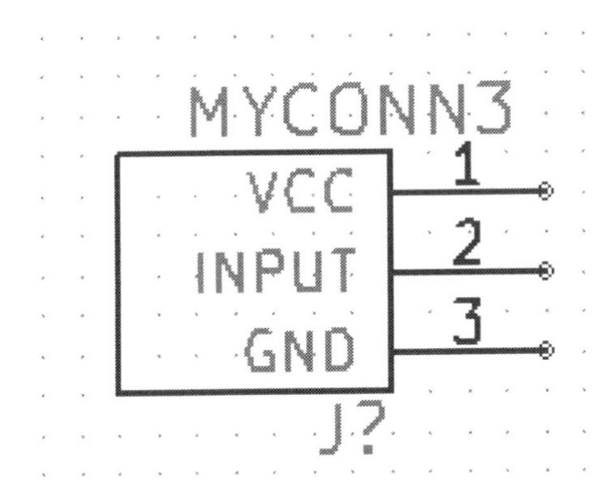

26. It is time to place the power and ground symbols. Click on the *Place a power port* button on the right toolbar. Alternatively, press the *p* key. In the component selection window, scroll down and select *VCC* from the *power* library. Click OK.

27. Click above the pin of the 1k resistor to place the VCC part. Click on the area above the microcontroller *VDD*. In the *Component Selection history* section select *VCC* and place it next to the VDD pin. Repeat the add process again and place a VCC part above the VCC pin of *MYCONN3*.

28. Repeat the add-pin steps but this time select the GND part. Place a GND part under the GND pin of *MYCONN3*. Place another GND symbol on the right of the VSS pin of the microcontroller. Your schematic should now look something like this:

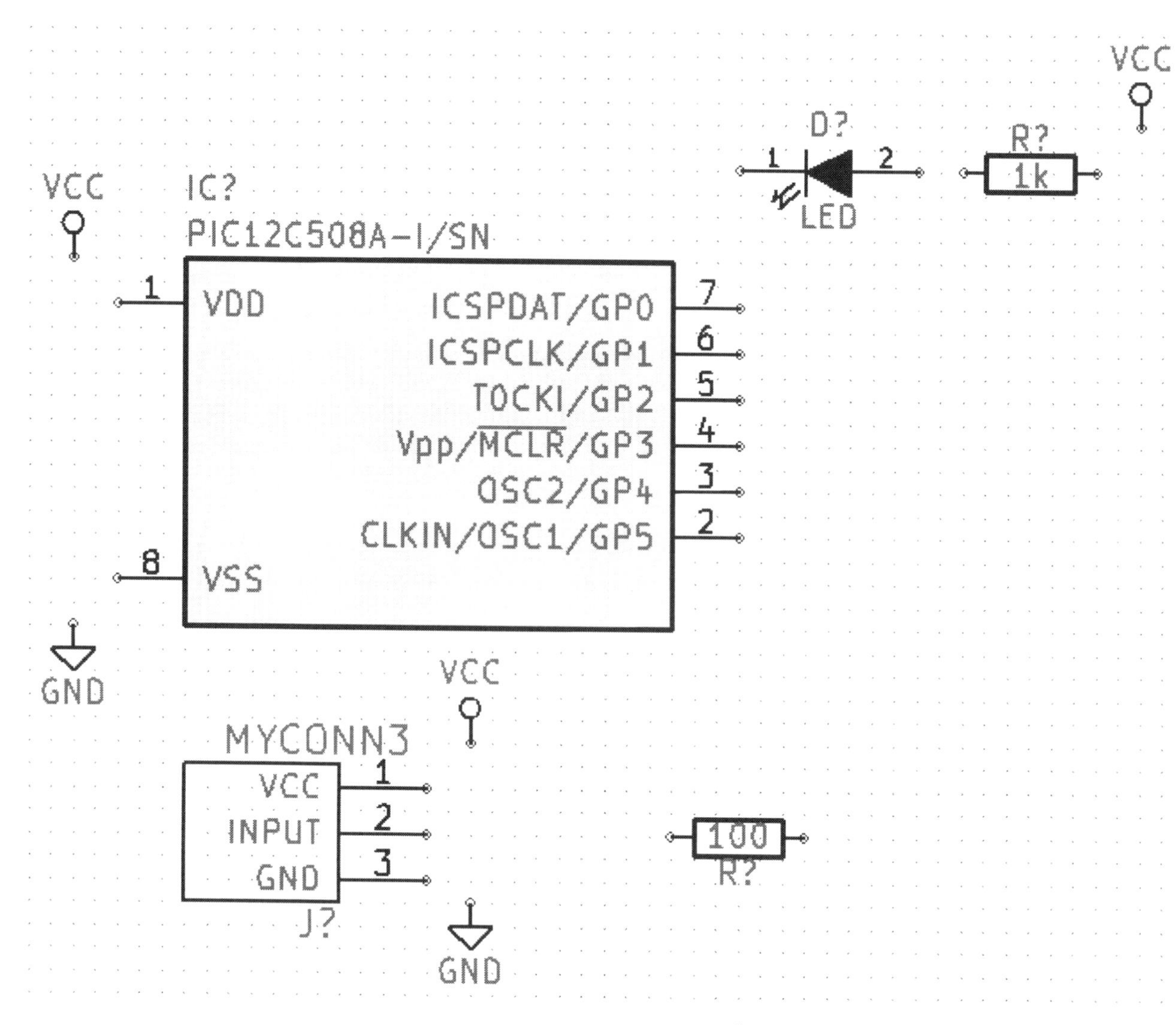

29. Next, we will wire all our components. Click on the *Place wire* icon on the right toolbar.

> **Note**
>
> Be careful not to pick *Place a bus*, which appears directly beneath this button but has thicker lines. The section Bus Connections in KiCad will explain how to use a bus section.

30. Click on the little circle at the end of pin 7 of the microcontroller and then click on the little circle on pin 2 of the LED. You can zoom in while you are placing the connection.

> **Note**
>
> If you want to reposition wired components, it is important to use the g key (grab) option and not the m key (move) option. Using the grab option will keep the wires connected. Review step 24 in case you have forgotten how to move a component.

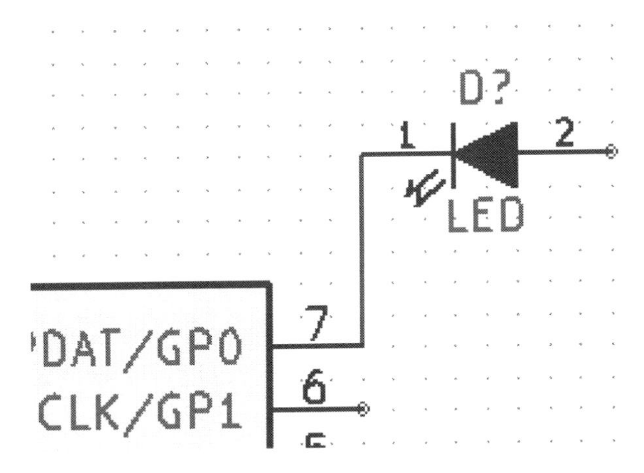

31. Repeat this process and wire up all the other components as shown below. To terminate a wire just double-click. When wiring up the VCC and GND symbols, the wire should touch the bottom of the VCC symbol and the middle top of the GND symbol. See the screenshot below.

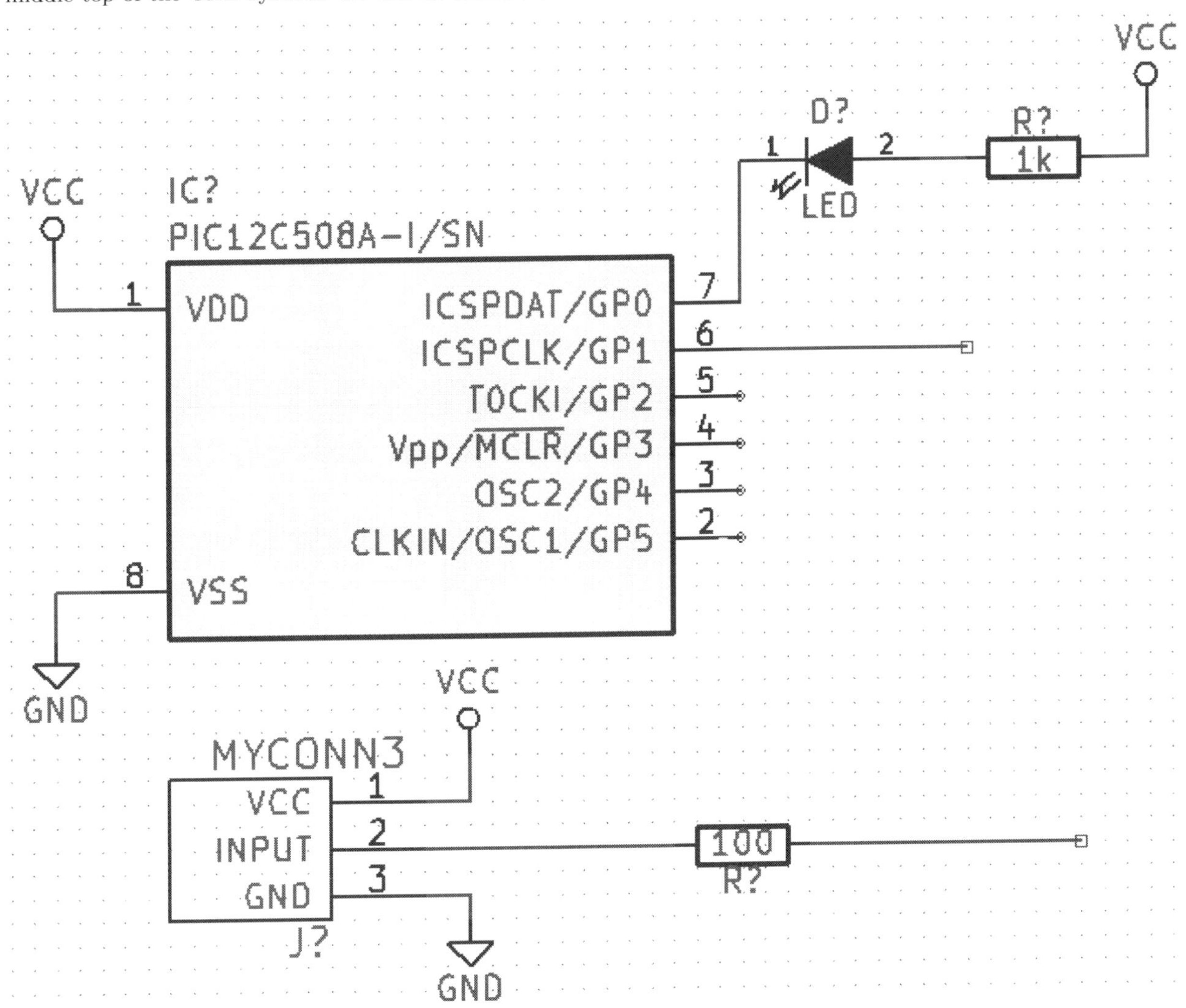

32. We will now consider an alternative way of making a connection using labels. Pick a net labelling tool by clicking

on the *Place net name* icon **A** on the right toolbar. You can also use the l key.

33. Click in the middle of the wire connected to pin 6 of the microcontroller. Name this label *INPUT*.

34. Follow the same procedure and place another label on line on the right of the 100 ohm resistor. Also name it *INPUT*. The two labels, having the same name, create an invisible connection between pin 6 of the PIC and the 100 ohm resistor. This is a useful technique when connecting wires in a complex design where drawing the lines would make the whole schematic messier. To place a label you do not necessarily need a wire, you can simply attach the label to a pin.

35. Labels can also be used to simply label wires for informative purposes. Place a label on pin 7 of the PIC. Enter the name *uCtoLED*. Name the wire between the resistor and the LED as *LEDtoR*. Name the wire between *MYCONN3* and the resistor as *INPUTtoR*.

36. You do not have to label the VCC and GND lines because the labels are implied from the power objects they are connected to.

37. Below you can see what the final result should look like.

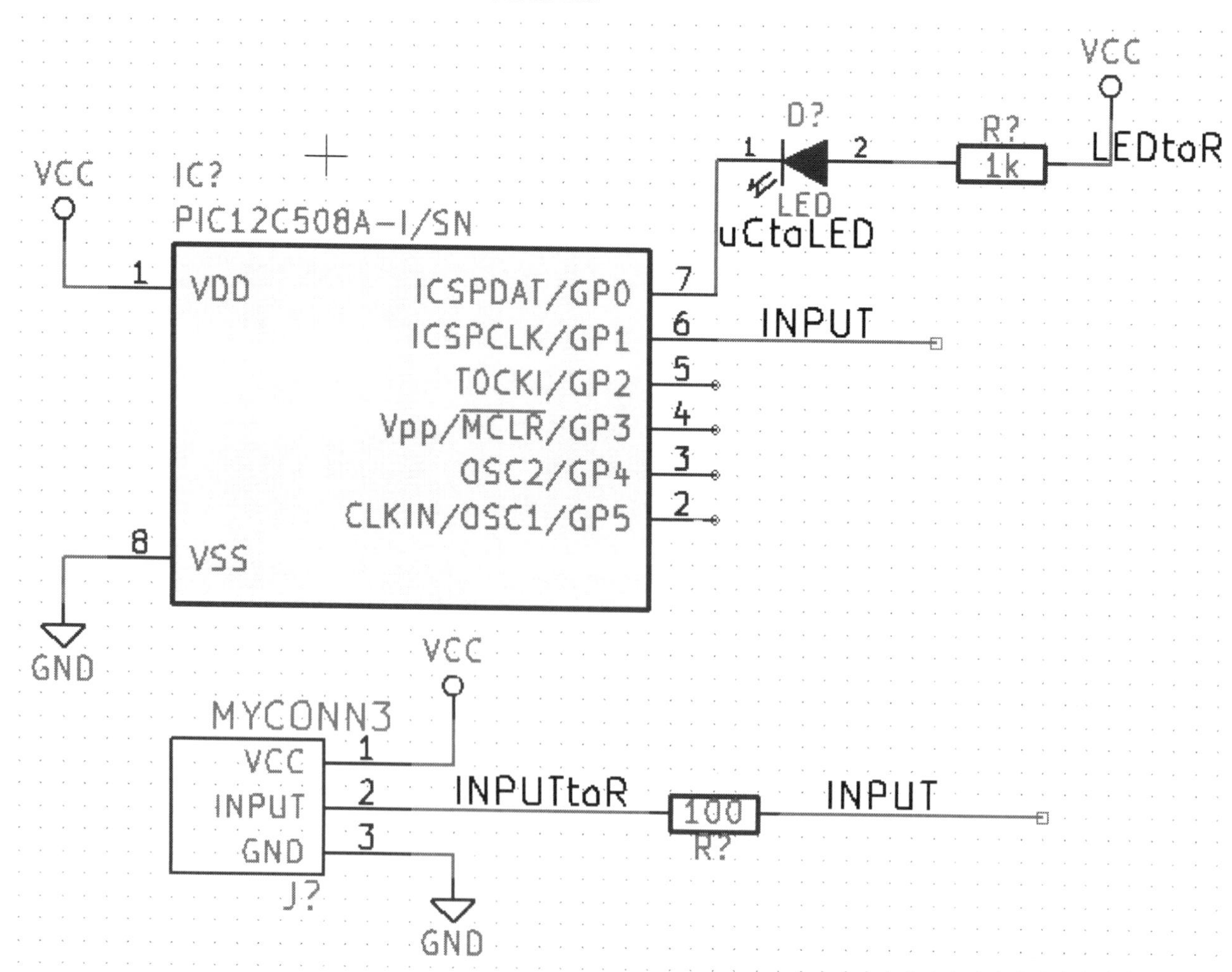

38. Let' s now deal with unconnected wires. Any pin or wire that is not connected will generate a warning when

checked by KiCad. To avoid these warnings you can either instruct the program that the unconnected wires are deliberate or manually flag each unconnected wire or pin as unconnected.

39. Click on the *Place no connect flag* icon on the right toolbar. Click on pins 2, 3, 4 and 5. An X will appear to signify that the lack of a wire connection is intentional.

40. Some components have power pins that are invisible. You can make them visible by clicking on the *Show hidden pins* icon on the left toolbar. Hidden power pins get automatically connected if VCC and GND naming is respected. Generally speaking, you should try not to make hidden power pins.

41. It is now necessary to add a *Power Flag* to indicate to KiCad that power comes in from somewhere. Press the a key, select *List All*, double click on the *power* library and search for *PWR_FLAG*. Place two of them. Connect them to a GND pin and to VCC as shown below.

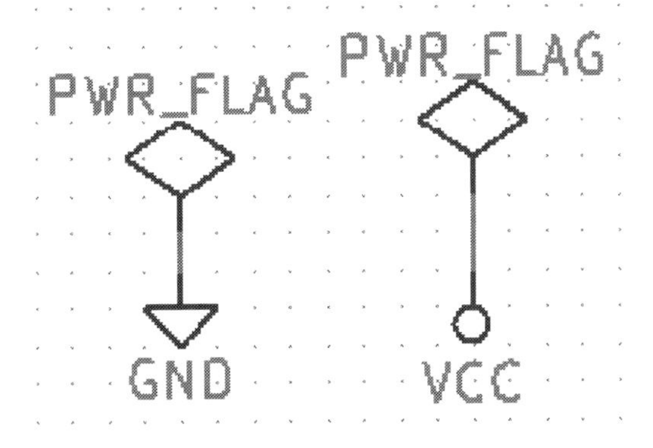

Note

This will avoid the classic schematic checking warning: Warning Pin power_in not driven (Net xx)

42. Sometimes it is good to write comments here and there. To add comments on the schematic use the *Place graphic text (comment)* icon on the right toolbar.

43. All components now need to have unique identifiers. In fact, many of our components are still named *R?* or *J?*. Identifier assignation can be done automatically by clicking on the *Annotate schematic* icon.

44. In the Annotate Schematic window, select *Use the entire schematic* and click on the *Annotation* button. Click OK in the confirmation message and then click *Close*. Notice how all the *?* have been replaced with numbers. Each identifier is now unique. In our example, they have been named *R1*, *R2*, *U1*, *D1* and *J1*.

45. We will now check our schematic for errors. Click on the *Perform Electric Rules Check* icon  . Click on the *Test ERC* button. A report informing you of any errors or warnings such as disconnected wires is generated. You should have 0 Errors and 0 Warnings. In case of errors or warnings, a small green arrow will appear on the schematic in the position where the error or the warning is located. Check *Write ERC report* and press the *Test ERC* button again to receive more information about the errors.

46. The schematic is now finished. We can now create a Netlist file to which we will add the footprint of each component. Click on the *Netlist generation* icon  on the top toolbar. Click on *Netlist* then click on *save*. Save under the default file name.

47. After generating the Netlist file, click on the *Run Cvpcb* icon  on the top toolbar. If a missing file error window pops up, just ignore it and click OK.

48. *Cvpcb* allows you to link all the components in your schematic with footprints in the KiCad library. The pane on the center shows all the components used in your schematic. Here select *D1*. In the pane on the right you have all the available footprints, here scroll down to *LEDs:LED-5MM* and double click on it.

49. It is possible that the pane on the right shows only a selected subgroup of available footprints. This is because KiCad is trying to suggest to you a subset of suitable footprints. Click on the icons 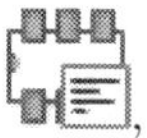 , 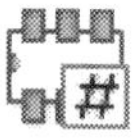 and 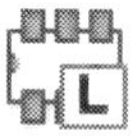 to enable or disable these filters.

50. For *IC1* select the *Housings_DIP:DIP-8_W7.62mm* footprint. For *J1* select the *Connect:Banana_Jack_3Pin* footprint. For *R1* and *R2* select the *Discret:R1* footprint.

51. If you are interested in knowing what the footprint you are choosing looks like, you have two options. You can click on the *View selected footprint* icon  for a preview of the current footprint. Alternatively, click on the *Display footprint list documentation* icon  and you will get a multi-page PDF document with all available footprints. You can print it out and check your components to make sure that the dimensions match.

52. You are done. You can now update your netlist file with all the associated footprints. Click on **File → Save As**. The default name *tutorial1.net* is fine, click save. Otherwise you can use the icon  . Your netlist file has now been updated with all the footprints. Note that if you are missing the footprint of any device, you will need to make your own footprints. This will be explained in a later section of this document.

53. You can close *Cvpcb* and go back to the *Eeschema* schematic editor. Save the project by clicking on **File → Save Whole Schematic Project**. Close the schematic editor.

54. Switch to the KiCad project manager.

55. The netlist file describes all components and their respective pin connections. The netlist file is actually a text file that you can easily inspect, edit or script.

Note

Library files (**.lib*) are text files too and they are also easily editable or scriptable.

56. To create a Bill Of Materials (BOM), go to the *Eeschema* schematic editor and click on the *Bill of materials* icon BOM on the top toolbar. By default there is no plugin active. You add one, by clicking on **Add Plugin** button. Select the *.xsl file you want to use, in this case, we select *bom2csv.xsl*.

Note

The *.xsl file is located in *plugins* directory of the KiCad installation, which is located at: /usr/lib/kicad/plugins/.

Or get the file via:

```
wget https://raw.githubusercontent.com/KiCad/kicad-source-mirror/master/eeschema/ ↩
    plugins/bom2csv.xsl
```

KiCad automatically generates the command, for example:

```
xsltproc -o "%O" "/home/<user>/kicad/eeschema/plugins/bom2csv.xsl" "%I"
```

You may want to add the extension, so change this command line to:

```
xsltproc -o "%O.csv" "/home/<user>/kicad/eeschema/plugins/bom2csv.xsl" "%I"
```

Press Help button for more info.

57. Now press *Generate*. The file (same name as your project) is located in your project folder. Open the ***.csv** file with LibreOffice Calc or Excel. An import window will appear, press OK.

You are now ready to move to the PCB layout part, which is presented in the next section. However, before moving on let's take a quick look at how to connect component pins using a bus line.

3.2 Bus connections in KiCad

Sometimes you might need to connect several sequential pins of component A with some other sequential pins of component B. In this case you have two options: the labelling method we already saw or the use of a bus connection. Let's see how to do it.

1. Let us suppose that you have three 4-pin connectors that you want to connect together pin to pin. Use the label option (press the l key) to label pin 4 of the P4 part. Name this label *a1*. Now let' s press the Ins key to have the same item automatically added on the pin below pin 4 (PIN 3). Notice how the label is automatically renamed *a2*.

2. Press the Ins Key two more times. The Ins key corresponds to the action *Repeat last item* and it is an infinitely useful command that can make your life a lot easier.

3. Repeat the same labelling action on the two other connectors CONN_2 and CONN_3 and you are done. If you proceed and make a PCB you will see that the three connectors are connected to each other. Figure 2 shows the result of what we described. For aesthetic purposes it is also possible to add a series of *Place wire to bus entry* using the icon and bus line using the icon , as shown in Figure 3. Mind, however, that there will be no effect on the PCB.

4. It should be pointed out that the short wire attached to the pins in Figure 2 is not strictly necessary. In fact, the labels could have been applied directly to the pins.

5. Let' s take it one step further and suppose that you have a fourth connector named CONN_4 and, for whatever reason, its labelling happens to be a little different (b1, b2, b3, b4). Now we want to connect *Bus a* with *Bus b* in a pin to pin manner. We want to do that without using pin labelling (which is also possible) but by instead using labelling on the bus line, with one label per bus.

6. Connect and label CONN_4 using the labelling method explained before. Name the pins b1, b2, b3 and b4. Connect the pin to a series of *Wire to bus entry* using the icon and to a bus line using the icon . See Figure 4.

7. Put a label (press the l key option) on the bus of CONN_4 and name it *b[1..4]*.

8. Put a label (press the l key option) on the previous a bus and name it *a[1..4]*.

9. What we can now do is connect bus a[1..4] with bus b[1..4] using a bus line with the button .

10. By connecting the two buses together, pin a1 will be automatically connected to pin b1, a2 will be connected to b2 and so on. Figure 4 shows what the final result looks like.

Note

The *Repeat last item* option accessible via the Ins key can be successfully used to repeat period item insertions. For instance, the short wires connected to all pins in Figure 2, Figure 3 and Figure 4 have been placed with this option.

---

11. The *Repeat last item* option accessible via the Ins key has also been extensively used to place the many series of *Wire to bus entry* using the icon .

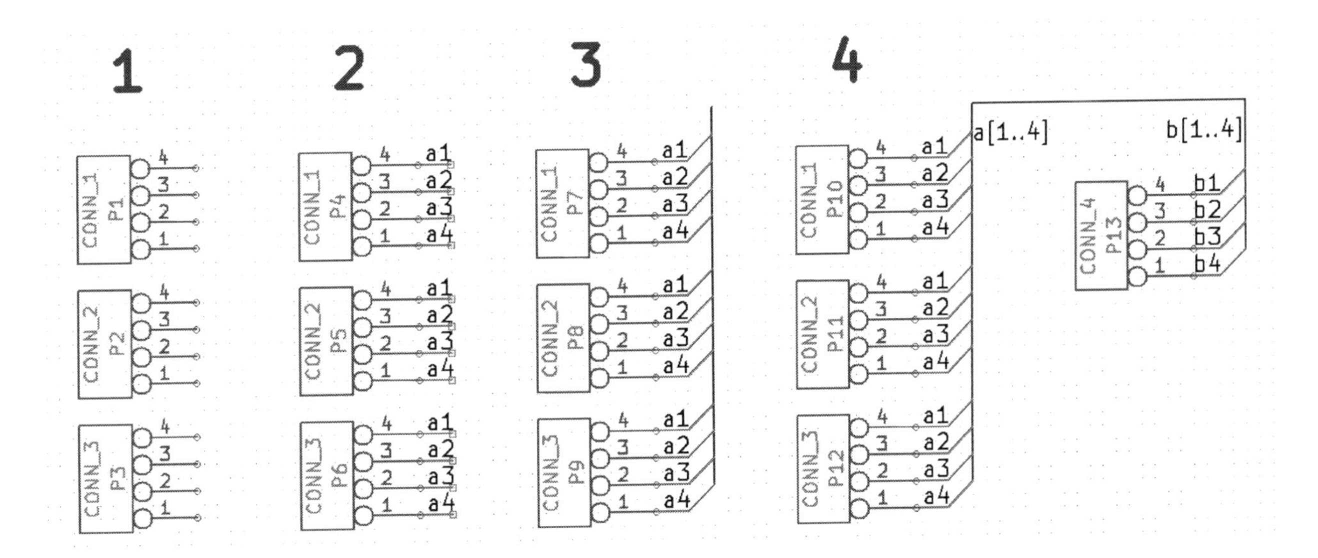
1
2
3
4
CONN_1
P1
CONN_2
P2
CONN_3
P3
CONN_1
P4
CONN_2
P5
CONN_3
P6
CONN_1
P7
CONN_2
P8
CONN_3
P9
CONN_1
P10
CONN_2
P11
CONN_3
P12
CONN_4
P13
a1
a2
a3
a4
a[1..4]
b[1..4]
b1
b2
b3
b4

Chapter 4

Layout printed circuit boards

It is now time to use the netlist file you generated to lay out the PCB. This is done with the *Pcbnew* tool.

4.1 Using Pcbnew

1. From the KiCad project manager, click on the *Pcbnew* icon . The *Pcbnew* window will open. If you get an error message saying that a **.kicad_pcb* file does not exist and asks if you want to create it, just click Yes.

2. Begin by entering some schematic information. Click on the *Page settings* icon on the top toolbar. Set *paper size* as *A4* and *title* as *Tutorial1*.

3. It is a good idea to start by setting the **clearance** and the **minimum track width** to those required by your PCB manufacturer. In general you can set the clearance to *0.25* and the minimum track width to *0.25*. Click on the **Design Rules** → **Design Rules** menu. If it does not show already, click on the *Net Classes Editor* tab. Change the *Clearance* field at the top of the window to *0.25* and the *Track Width* field to *0.25* as shown below. Measurements here are in mm.

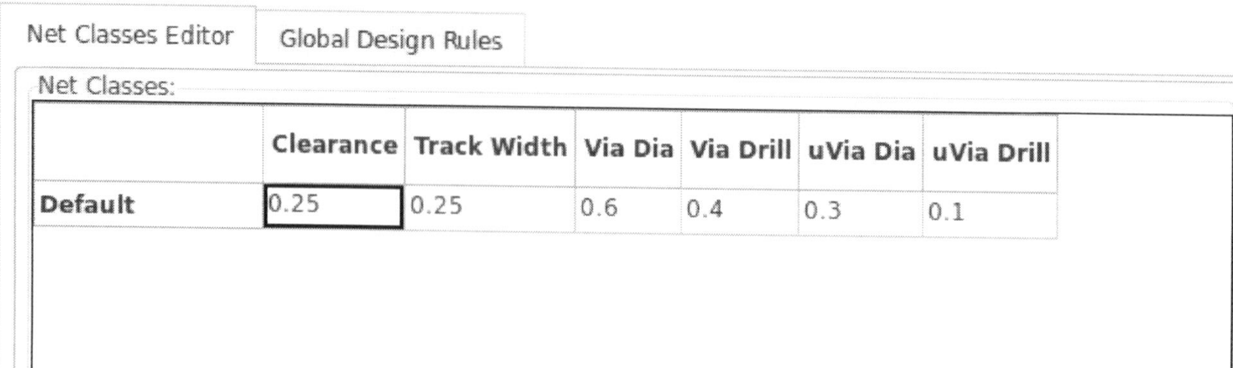

	Clearance	Track Width	Via Dia	Via Drill	uVia Dia	uVia Drill
Default	0.25	0.25	0.6	0.4	0.3	0.1

4. Click on the *Global Design Rules* tab and set *Min track width* to 0.25'. Click the OK button to commit your changes and close the Design Rules Editor window.

5. Now we will import the netlist file. Click on the *Read Netlist* icon on the top toolbar. Click on the *Browse Netlist Files* button, select *tutorial1.net* in the File selection dialogue, and click on *Read Current Netlist*. Then click the *Close* button.

6. All components should now be visible in the top left hand corner just above the page. Scroll up if you cannot see them.

7. Select all components with the mouse and move them to the middle of the board. If necessary you can zoom in and out while you move the components.

8. All components are connected via a thin group of wires called *ratsnest*. Make sure that the *Hide board ratsnest* button is pressed. In this way you can see the ratsnest linking all components.

Note

The tool-tip is backwards; pressing this button actually displays the ratsnest.

9. You can move each component by hovering over it and pressing the g key. Click where you want to place them. Move all components around until you minimise the number of wire crossovers.

Note

If instead of grabbing the components (with the g key) you move them around using the m key you will later note that you lose the track connection (the same occurs in the schematic editor). Bottom line, always use the g key option.

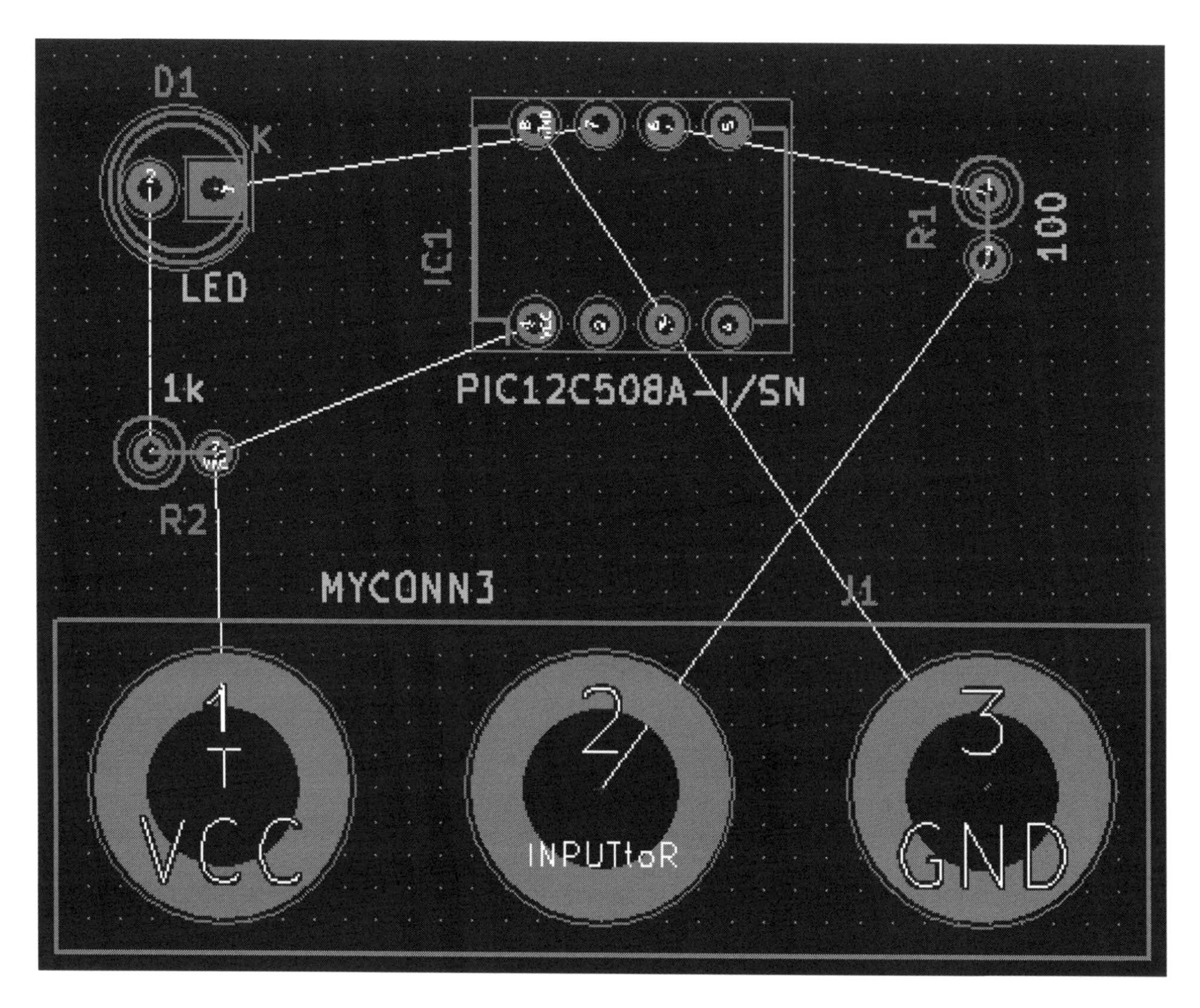

10. If the ratsnest disappears or the screen gets messy, right click and click *Redraw view*. Note how one pin of the 100 ohm resistor is connected to pin 6 of the PIC component. This is the result of the labelling method used to connect pins. Labels are often preferred to actual wires because they make the schematic much less messy.

11. Now we will define the edge of the PCB. Select *Edge.Cuts* from the drop down menu in the top toolbar. Click on the *Add graphic line or polygon* icon on the right toolbar. Trace around the edge of the board, clicking at each corner, and remember to leave a small gap between the edge of the green and the edge of the PCB.

12. Next, connect up all the wires except GND. In fact, we will connect all GND connections in one go using a ground plane placed on the bottom copper (called *B.Cu*) of the board.

13. Now we must choose which copper layer we want to work on. Select *F.Cu (PgUp)* in the drag down menu on the top toolbar. This is the front top copper layer.

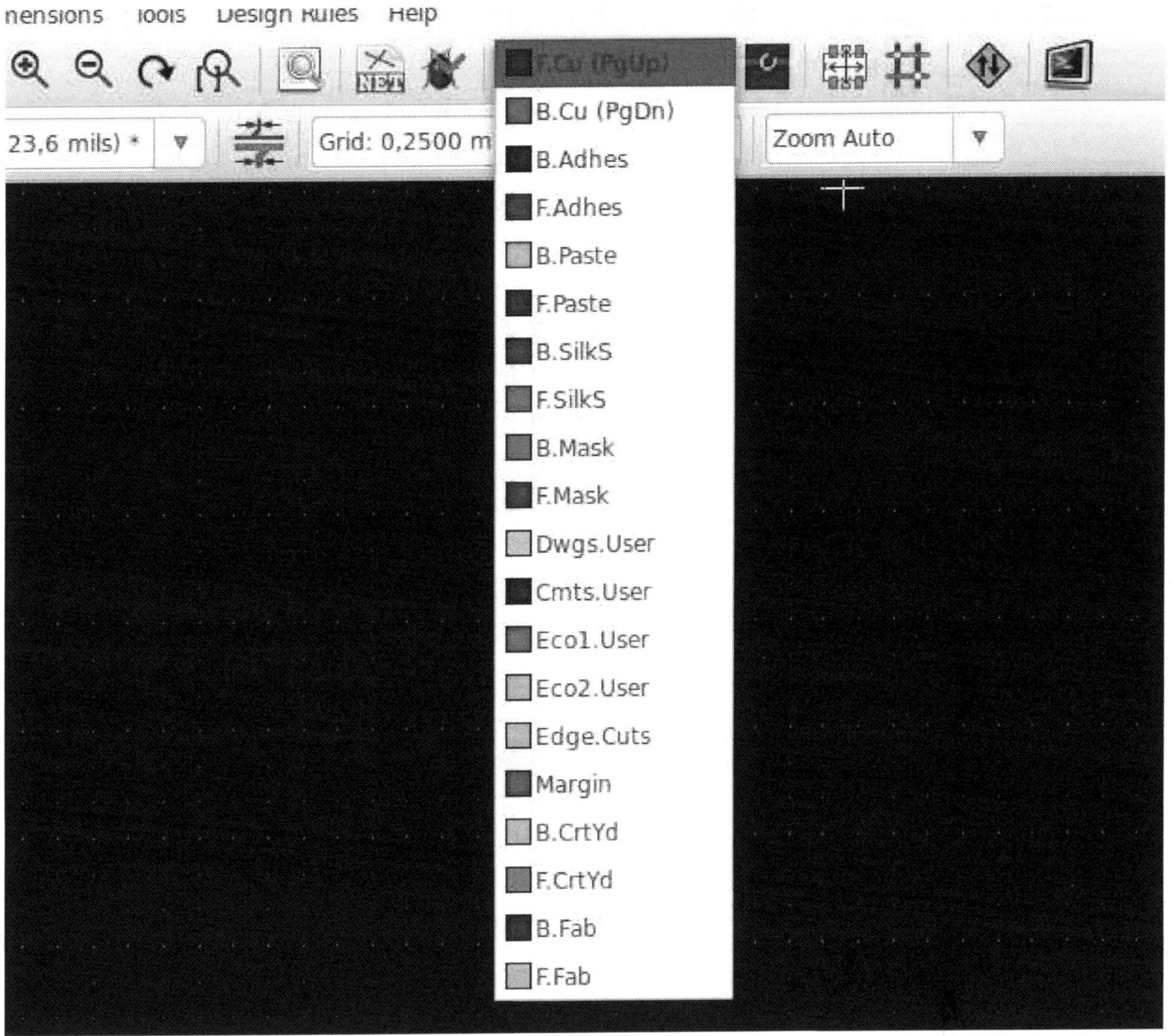

14. If you decide, for instance, to do a 4 layer PCB instead, go to **Design Rules** → **Layers Setup** and change *Copper Layers* to 4. In the *Layers* table you can name layers and decide what they can be used for. Notice that there are very useful presets that can be selected via the *Preset Layer Groupings* menu.

15. Click on the *Add Tracks and vias* icon on the right toolbar. Click on pin 1 of *J1* and run a track to pad *R2*. Double-click to set the point where the track will end. The width of this track will be the default 0.250 mm. You can change the track width from the drop-down menu in the top toolbar. Mind that by default you have only one track width available.

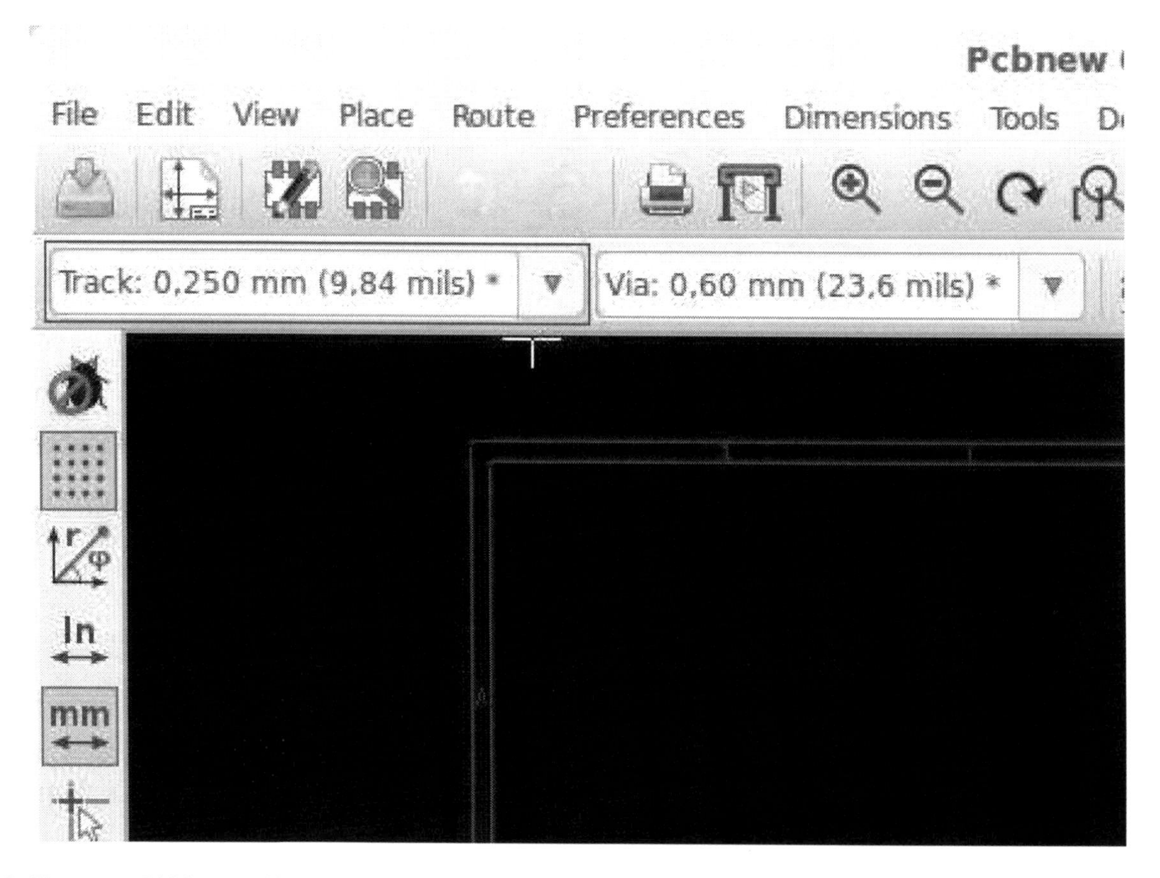

16. If you would like to add more track widths g o to: **Design Rules** → **Design Rules** → **Global Design Rules** tab and at the bottom right of this window add any other width you would like to have available. You can then choose the widths of the track from the drop-down menu while you lay out your board. See the example below (inches).

Custom Track Widths:

	Width
Track 1	0.0100
Track 2	0.0200
Track 3	0.0500
Track 4	0.0800
Track 5	0.1000
Track 6	0.1500
Track 7	0.2000

17. Alternatively, you can add a Net Class in which you specify a set of options. Go to **Design Rules → Design Rules → Net Classes Editor** and add a new class called *power*. Change the track width from 8 mil (indicated as 0.0080) to 24 mil (indicated as 0.0240). Next, add everything but ground to the *power* class (select *default* at left and *power* at right and use the arrows).

18. If you want to change the grid size, **Right click → Grid Select**. Be sure to select the appropriate grid size before or after laying down the components and connecting them together with tracks.

19. Repeat this process until all wires, except pin 3 of J1, are connected. Your board should look like the example below.

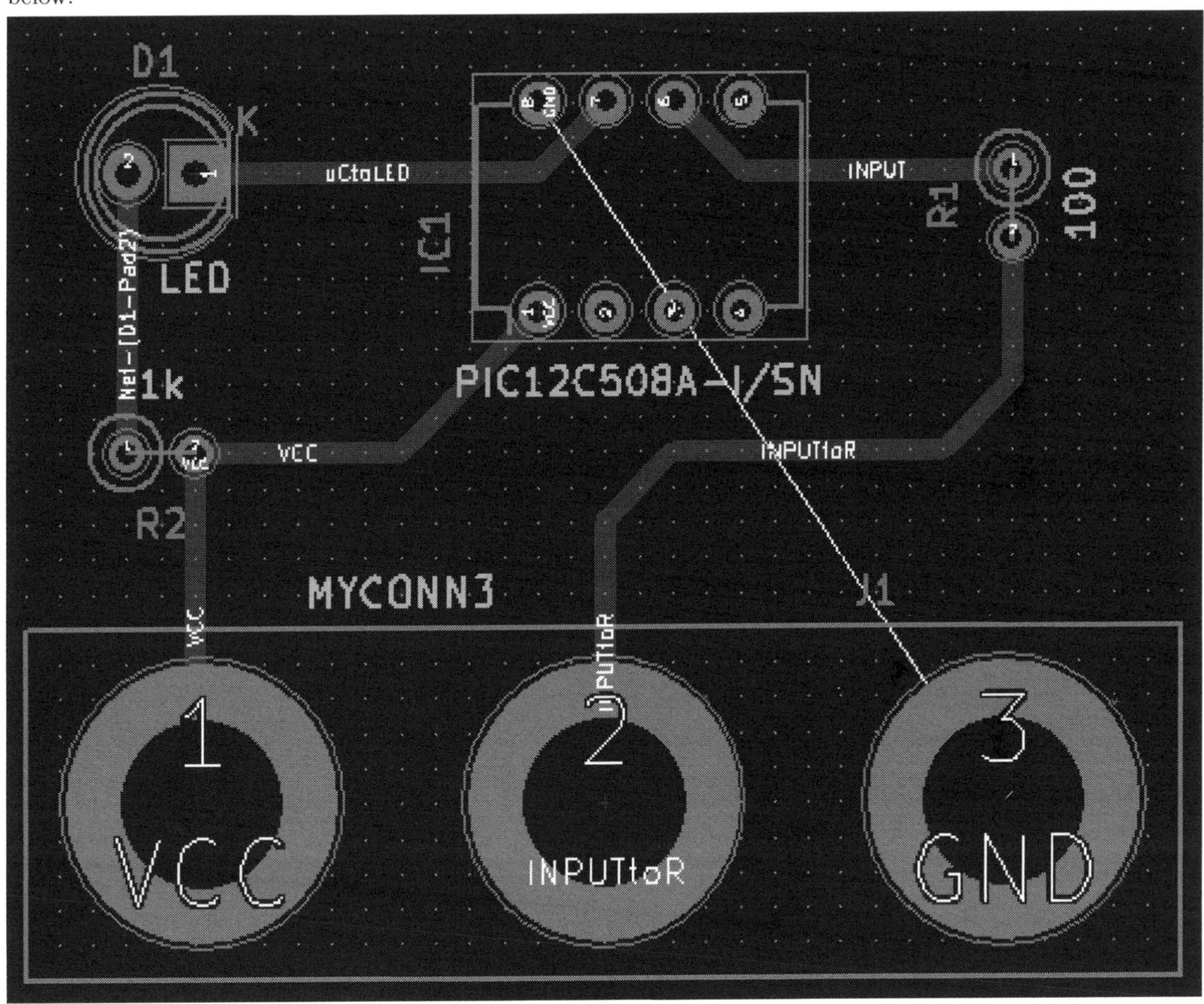

20. Let' s now run a track on the other copper side of the PCB. Select *B.Cu* in the drag down menu on the top toolbar. Click on the *Add tracks and vias* icon . Draw a track between pin 3 of J1 and pin 8 of U1. This is actually not necessary since we could do this with the ground plane. Notice how the colour of the track has changed.

21. **Go from pin A to pin B by changing layer**. It is possible to change the copper plane while you are running a track by placing a via. While you are running a track on the upper copper plane, right click and select *Place Via* or simply press the v key. This will take you to the bottom layer where you can complete your track.

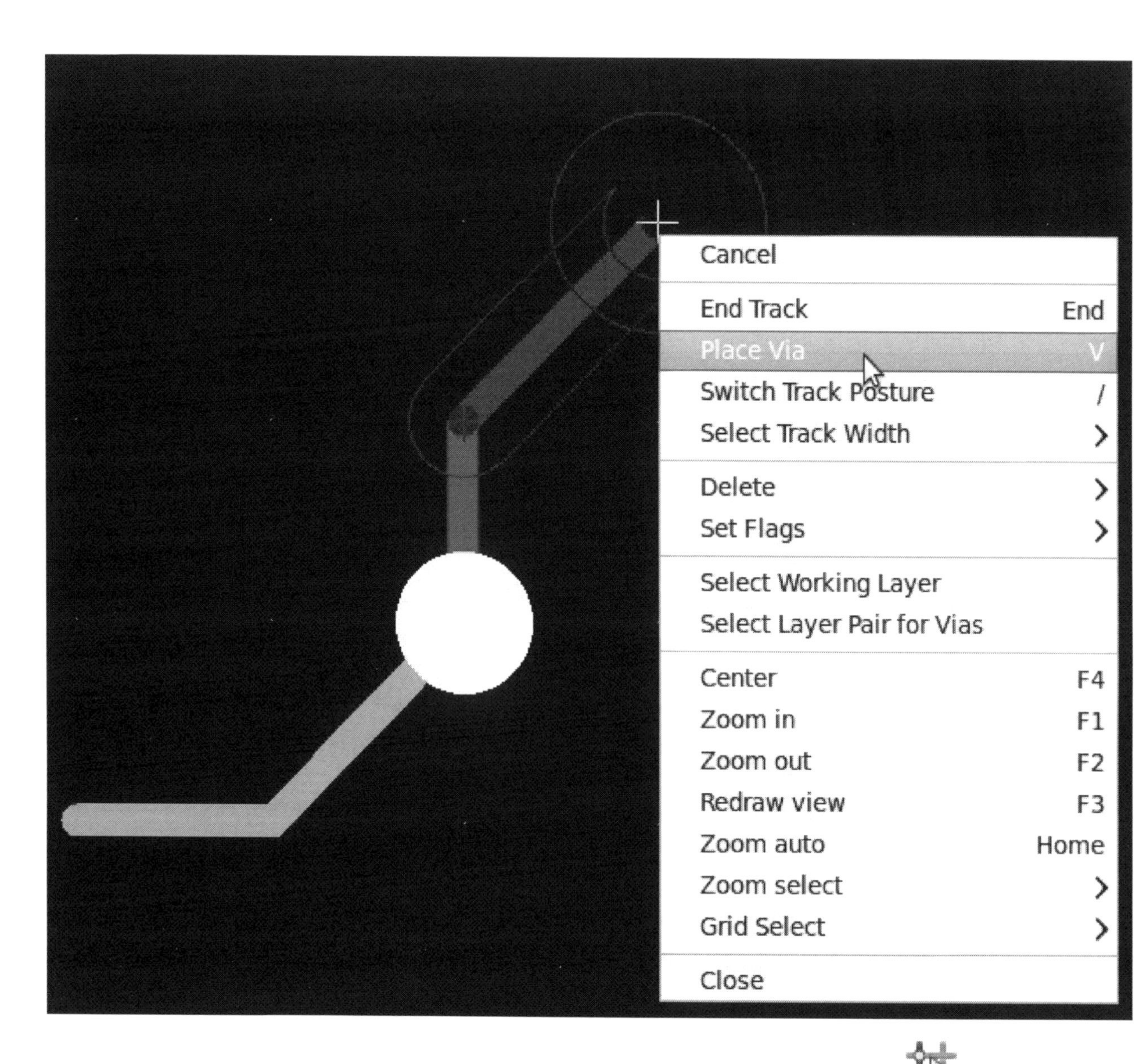

22. When you want to inspect a particular connection you can click on the *Net highlight* icon on the right toolbar. Click on pin 3 of J1. The track itself and all pads connected to it should become highlighted.

23. Now we will make a ground plane that will be connected to all GND pins. Click on the *Add Zones* icon on the right toolbar. We are going to trace a rectangle around the board, so click where you want one of the corners to be. In the dialogue that appears, set *Pad in Zone* to *Thermal relief* and *Zone edges orient* to *H,V* and click OK.

24. Trace around the outline of the board by clicking each corner in rotation. Double-click to finish your rectangle. Right click inside the area you have just traced. Click on *Fill or Refill All Zones*. The board should fill in with green and look something like this:

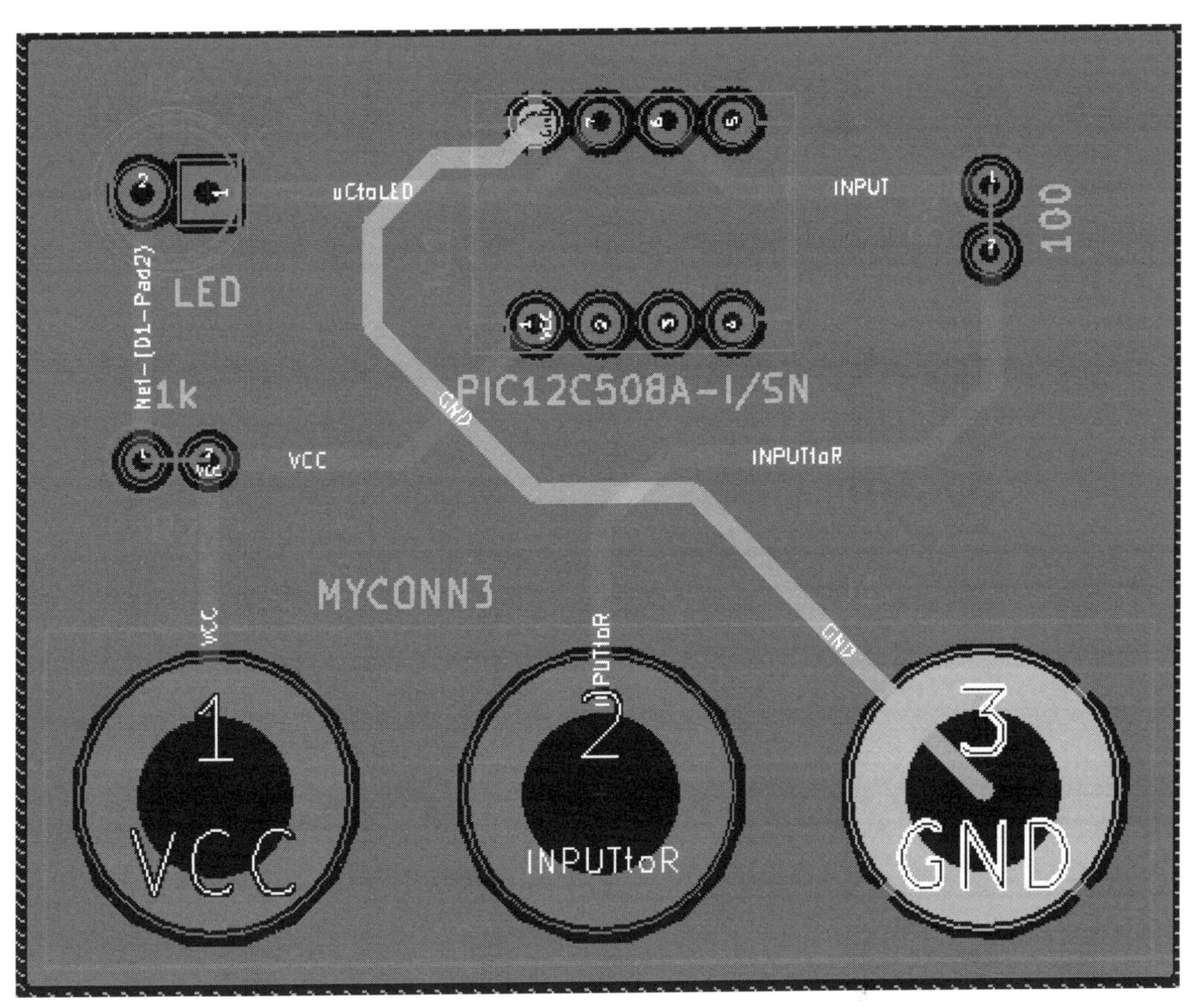

25. Run the design rules checker by clicking on the *Perform Design Rules Check* icon on the top toolbar. Click on *Start DRC*. There should be no errors. Click on *List Unconnected*. There should be no unconnected track. Click OK to close the DRC Control dialogue.

26. Save your file by clicking on **File** → **Save**. To admire your board in 3D, click on **View** → **3D Viewer**.

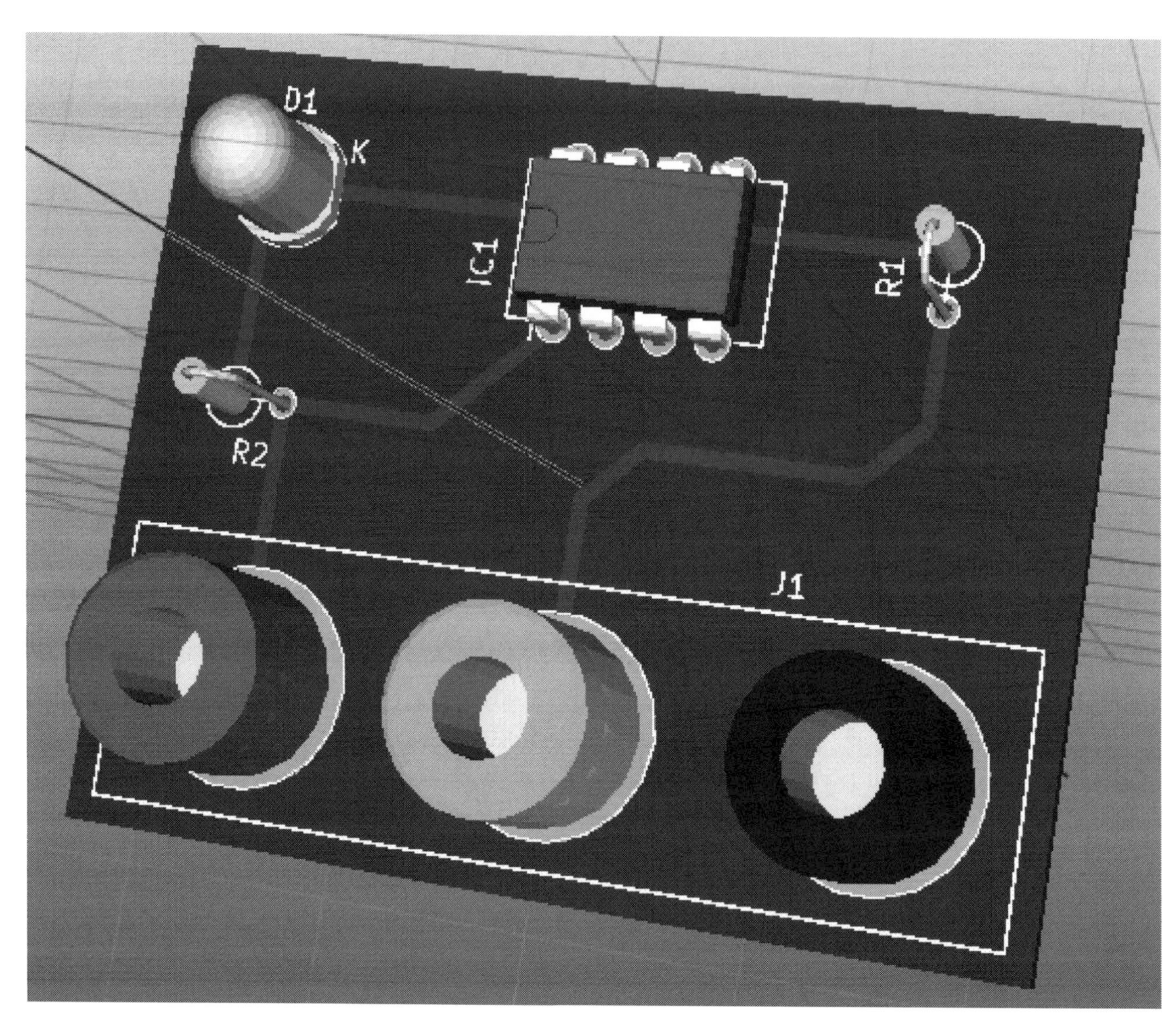

27. You can drag your mouse around to rotate the PCB.
28. Your board is complete. To send it off to a manufacturer you will need to generate all Gerber files.

4.2 Generate Gerber files

Once your PCB is complete, you can generate Gerber files for each layer and send them to your favourite PCB manufacturer, who will make the board for you.

1. From KiCad, open the *Pcbnew* software tool and load your board file by clicking on the icon 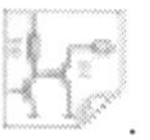.
2. Click on **File** → **Plot**. Select *Gerber* as the *Plot Format* and select the folder in which to put all Gerber files. Proceed by clicking on the *Plot* button.
3. These are the layers you need to select for making a typical 2-layer PCB:

Layer	KiCad Layer Name	Old KiCad Layer Name	Default Gerber Extension	"Use Protel filename extensions" is enabled
Bottom Layer	B.Cu	Copper	.GBR	.GBL
Top Layer	F.Cu	Component	.GBR	.GTL
Top Overlay	F.SilkS	SilkS_Cmp	.GBR	.GTO
Bottom Solder Resist	B.Mask	Mask_Cop	.GBR	.GBS
Top Solder Resist	F.Mask	Mask_Cmp	.GBR	.GTS
Edges	Edge.Cuts	Edges_Pcb	.GBR	.GM1

4.3 Using GerbView

1. To view all your Gerber files go to the KiCad project manager and click on the *GerbView* icon. On the drag down menu select *Layer 1*. Click on **File → Load Gerber file** or click on the icon . Load all generated Gerber files one at a time. Note how they all get displayed one on top of the other.
2. Use the menu on the right to select/deselect which layer to show. Carefully inspect each layer before sending them for production.
3. To generate the drill file, from *Pcbnew* go again for the **File → Plot** option. Default settings should be fine.

4.4 Automatically route with FreeRouter

Routing a board by hand is quick and fun, however, for a board with lots of components you might want to use an autorouter. Remember that you should first route critical traces by hand and then set the autorouter to do the boring bits. Its work will only account for the unrouted traces. The autorouter we will use here is FreeRouter from *freerouting.net*.

Note
Freerouter is a open source java application, and it is needed to build by yourself to use with KiCad. Source code of Freerouter can be found on this site: https://github.com/nikropht/FreeRouting

---

1. From *Pcbnew* click on **File → Export → Specctra DSN** or click on **Tools → FreeRoute → Export a Specctra Design (*.dsn) file** and save the file locally. Launch FreeRouter and click on the *Open Your Own Design* button, browse for the *dsn* file and load it.

   Note
   The **Tools → FreeRoute** dialog has a nice help button that opens a file viewer with a little document inside named **Freerouter Guidelines**. Please follow these guidelines to use FreeRoute effectively.

   ---

2. FreeRouter has some features that KiCad does not currently have, both for manual routing and for automatic routing. FreeRouter operates in two main steps: first, routing the board and then optimising it. Full optimisation can take a long time, however you can stop it at any time need be.

3. You can start the automatic routing by clicking on the *Autorouter* button on the top bar. The bottom bar gives you information about the on-going routing process. If the *Pass* count gets above 30, your board probably can not be autorouted with this router. Spread your components out more or rotate them better and try again. The goal in rotation and position of parts is to lower the number of crossed airlines in the ratsnest.

4. Making a left-click on the mouse can stop the automatic routing and automatically start the optimisation process. Another left-click will stop the optimisation process. Unless you really need to stop, it is better to let FreeRouter finish its job.

5. Click on the **File** → **Export Specctra Session File** menu and save the board file with the *.ses* extension. You do not really need to save the FreeRouter rules file.

6. Back to *Pcbnew*. You can import your freshly routed board by clicking on the link **Tools** → **FreeRoute** and then on the icon *Back Import the Spectra Session (.ses) File* and selecting your *.ses* file.

If there is any routed trace that you do not like, you can delete it and re-route it again, using the del key and the routing tool, which is the *Add tracks* icon on the right toolbar.

Chapter 5

Forward annotation in KiCad

Once you have completed your electronic schematic, the footprint assignment, the board layout and generated the Gerber files, you are ready to send everything to a PCB manufacturer so that your board can become reality.

Often, this linear work-flow turns out to be not so uni-directional. For instance, when you have to modify/extend a board for which you or others have already completed this work-flow, it is possible that you need to move components around, replace them with others, change footprints and much more. During this modification process, what you do not want to do is to re-route the whole board again from scratch. Instead, this is how you do it:

1. Let′ s suppose that you want to replace a hypothetical connector CON1 with CON2.
2. You already have a completed schematic and a fully routed PCB.
3. From KiCad, start *Eeschema*, make your modifications by deleting CON1 and adding CON2. Save your schematic project with the icon and c lick on the *Netlist generation* icon on the top toolbar.
4. Click on *Netlist* then on *save*. Save to the default file name. You have to rewrite the old one.
5. Now assign a footprint to CON2. Click on the *Run Cvpcb* icon on the top toolbar. Assign the footprint to the new device CON2. The rest of the components still have the previous footprints assigned to them. Close *Cvpcb*.
6. Back in the schematic editor, save the project by clicking on *File* → *Save Whole Schematic Project*. Close the schematic editor.
7. From the KiCad project manager, click on the *Pcbnew* icon. The *Pcbnew* window will open.
8. The old, already routed, board should automatically open. Let′ s import the new netlist file. Click on the *Read Netlist* icon on the top toolbar.
9. Click on the *Browse Netlist Files* button, select the netlist file in the file selection dialogue, and click on *Read Current Netlist*. Then click the *Close* button.

10. At this point you should be able to see a layout with all previous components already routed. On the top left corner you should see all unrouted components, in our case the CON2. Select CON2 with the mouse. Move the component to the middle of the board.

11. Place CON2 and route it. Once done, save and proceed with the Gerber file generation as usual.

The process described here can easily be repeated as many times as you need. Beside the Forward Annotation method described above, there is another method known as Backward Annotation. This method allows you to make modifications to your already routed PCB from Pcbnew and updates those modifications in your schematic and netlist file. The Backward Annotation method, however, is not that useful and is therefore not described here.

Chapter 6

Make schematic components in KiCad

Sometimes a component that you want to place on your schematic is not in the KiCad libraries. This is quite normal and there is no reason to worry. In this section we will see how a new schematic component can be quickly created with KiCad. Nevertheless, remember that you can always find KiCad components on the Internet. For instance from here:

http://per.launay.free.fr/kicad/kicad_php/composant.php

In KiCad, a component is a piece of text that starts with a *DEF* and ends with *ENDDEF*. One or more components are normally placed in a library file with the extension *.lib*. If you want to add components to a library file you can just use the cut and paste commands.

6.1 Using Component Library Editor

1. We can use the *Component Library Editor* (part of *Eeschema*) to make new components. In our project folder *tutorial1* let' s create a folder named *library*. Inside we will put our new library file *myLib.lib* as soon as we have created our new component.
2. Now we can start creating our new component. From KiCad, start *Eeschema*, click on the *Library Editor* icon and then click on the *New component* icon . The Component Properties window will appear. Name the new component *MYCONN3*, set the *Default reference designator* as *J*, and the *Number of parts per package* as *1*. Click OK. If the warning appears just click yes. At this point the component is only made of its labels.

   Let' s add some pins. Click on the *Add Pins* icon on the right toolbar. To place the pin, left click in the centre of the part editor sheet just below the *MYCONN3* label.
3. In the Pin Properties window that appears, set the pin name to *VCC*, set the pin number to *1*, and the *Electrical type* to *Passive* then click OK.

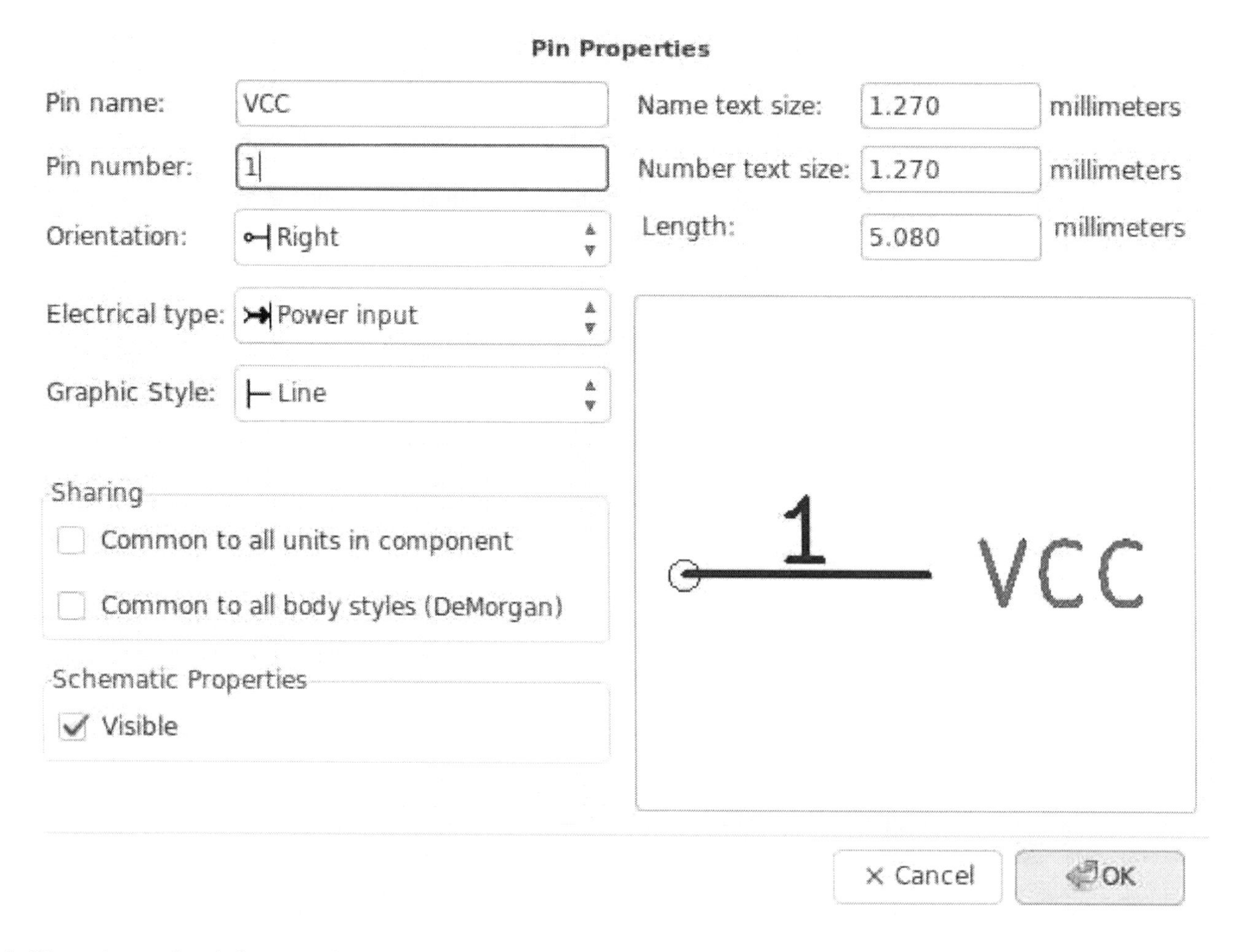

4. Place the pin by clicking on the location you would like it to go, right below the *MYCONN3* label.
5. Repeat the place-pin steps, this time *Pin name* should be *INPUT*, *Pin number* should be *2*, and *Electrical Type* should be *Passive*.
6. Repeat the place-pin steps, this time *Pin name* should be *GND*, *Pin number* should be *3*, and *Electrical Type* should be *Passive*. Arrange the pins one on top of the other. The component label *MYCONN3* should be in the centre of the page (where the blue lines cross).
7. Next, draw the contour of the component. Click on the *Add rectangle* icon . We want to draw a rectangle next to the pins, as shown below. To do this, click where you want the top left corner of the rectangle to be. Click again where you want the bottom right corner of the rectangle to be.

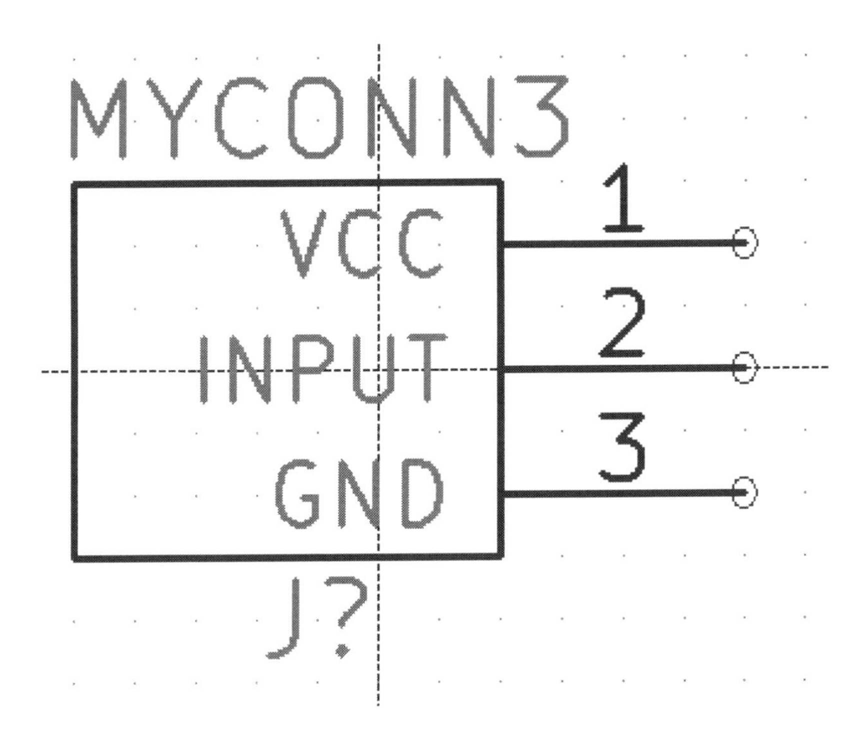

8. Save the component in your library *myLib.lib*. Click on the *New Library* icon , navigate into *tutorial1/ library/* folder and save the new library file with the name *myLib.lib*.
9. Go to **Preferences** → **Component Libraries** and add both *tutorial1/library/* in *User defined search path* and *myLib.lib in Component library files.*
10. Click on the *Select working library* icon . In the Select Library window click on *myLib* and click OK. Notice how the heading of the window indicates the library currently in use, which now should be *myLib*.
11. Click on the *Update current component in current library* icon in the top toolbar. Save all changes by clicking on the *Save current loaded library on disk* icon in the top toolbar. Click *Yes* in any confirmation messages that appear. The new schematic component is now done and available in the library indicated in the window title bar.
12. You can now close the Component library editor window. You will return to the schematic editor window. Your new component will now be available to you from the library *myLib*.
13. You can make any library *file.lib* file available to you by adding it to the library path. From *Eeschema*, go to **Preferences** → **Library** and add both the path to it in *User defined search path* and *file.lib* in *Component library files*.

6.2 Export, import and modify library components

Instead of creating a library component from scratch it is sometimes easier to start from one already made and modify it. In this section we will see how to export a component from the KiCad standard library *device* to your own library *myOwnLib.lib* and then modify it.

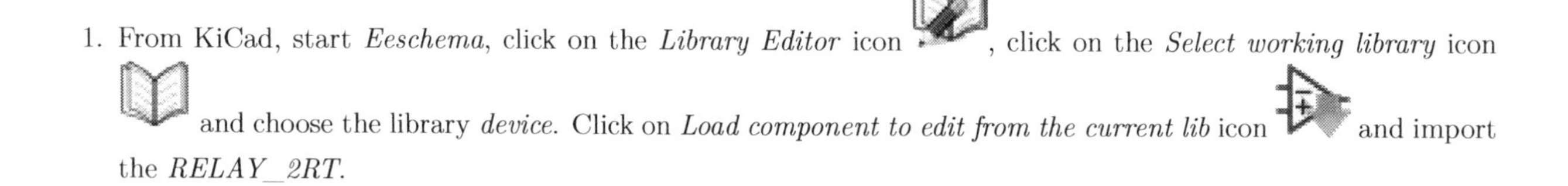

1. From KiCad, start *Eeschema*, click on the *Library Editor* icon, click on the *Select working library* icon and choose the library *device*. Click on *Load component to edit from the current lib* icon and import the *RELAY_2RT*.
2. Click on the *Export component* icon, navigate into the *library/* folder and save the new library file with the name *myOwnLib.lib*.
3. You can make this component and the whole library *myOwnLib.lib* available to you by adding it to the library path. From *Eeschema*, go to **Preferences** → **Component Libraries** and add both *library/* in *User defined search path* and *myOwnLib.lib* in the *Component library files*.
4. Click on the *Select working library* icon. In the Select Library window click on *myOwnLib* and click OK. Notice how the heading of the window indicates the library currently in use, it should be *myOwnLib*.
5. Click on the *Load component to edit from the current lib* icon and import the *RELAY_2RT*.
6. You can now modify the component as you like. Hover over the label *RELAY_2RT*, press the e key and rename it *MY_RELAY_2RT*.
7. Click on *Update current component in current library* icon in the top toolbar. Save all changes by clicking on the *Save current loaded library on disk* icon in the top toolbar.

6.3 Make schematic components with quicklib

This section presents an alternative way of creating the schematic component for MYCONN3 (see MYCONN3 above) using the Internet tool *quicklib*.

1. Head to the *quicklib* web page: http://kicad.rohrbacher.net/quicklib.php
2. Fill out the page with the following information: Component name: MYCONN3 Reference Prefix: J Pin Layout Style: SIL Pin Count, N: 3
3. Click on the *Assign Pins* icon. Fill out the page with the following information: Pin 1: VCC Pin 2: input Pin 3: GND. Type : Passive for all 3 pins.
4. Click on the icon *Preview it* and, if you are satisfied, click on the *Build Library Component*. Download the file and rename it *tutorial1/library/myQuickLib.lib.*. You are done!
5. Have a look at it using KiCad. From the KiCad project manager, start *Eeschema*, click on the *Library Editor* icon, click on the *Import Component* icon, navigate to *tutorial1/library/* and select *myQuickLib.lib*.

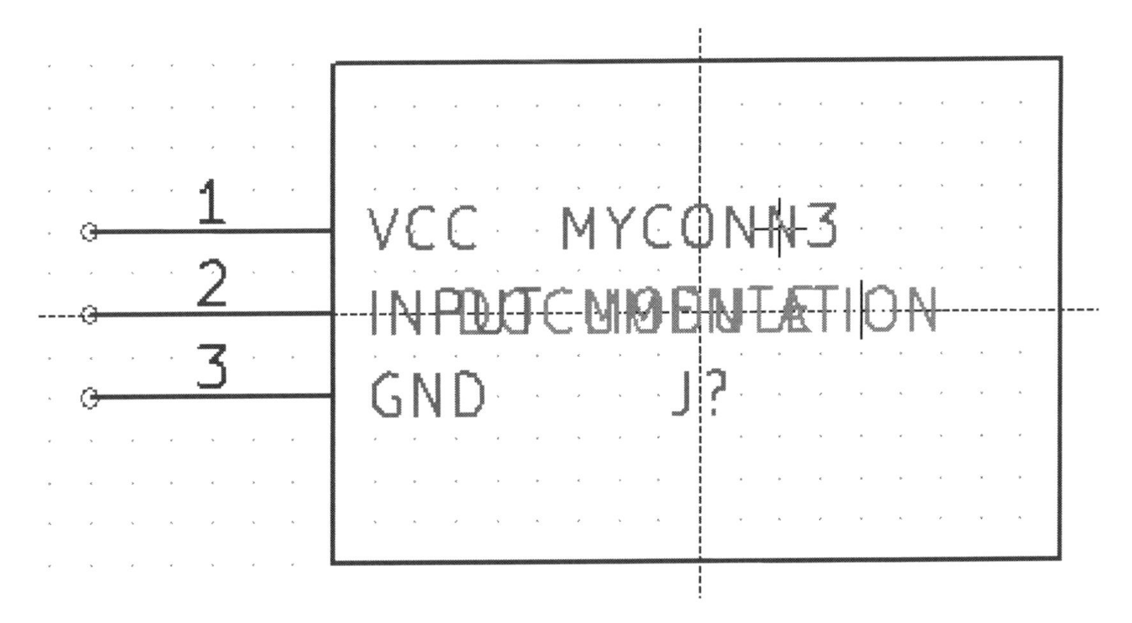

6. You can make this component and the whole library *myQuickLib.lib* available to you by adding it to the KiCad library path. From *Eeschema*, go to **Preferences** → **Component Libraries** and add *library* in *User defined search path* and *myQuickLib.lib* in *Component library files.*

As you might guess, this method of creating library components can be quite effective when you want to create components with a large pin count.

6.4 Make a high pin count schematic component

In the section titled *Make Schematic Components in quicklib* we saw how to make a schematic component using the *quicklib* web-based tool. However, you will occasionally find that you need to create a schematic component with a high number of pins (some hundreds of pins). In KiCad, this is not a very complicated task.

1. Suppose that you want to create a schematic component for a device with 50 pins. It is common practise to draw it using multiple low pin-count drawings, for example two drawings with 25 pins each. This component representation allows for easy pin connection.
2. The best way to create our component is to use *quicklib* to generate two 25-pin components separately, re-number their pins using a Python script and finally merge the two by using copy and paste to make them into one single DEF and ENDDEF component.
3. You will find an example of a simple Python script below that can be used in conjunction with an *in.txt* file and an *out.txt* file to re-number the line: `X PIN1 1 -750 600 300 R 50 50 1 1 I` into `X PIN26 26 -750 600 300 R 50 50 1 1 I` this is done for all lines in the file *in.txt*.

Simple script

```
#!/usr/bin/env python
''' simple script to manipulate KiCad component pins numbering'''
import sys, re
```

```
try:
    fin=open(sys.argv[1],'r')
    fout=open(sys.argv[2],'w')
except:
    print "oh, wrong use of this app, try:", sys.argv[0], "in.txt out.txt"
    sys.exit()
for ln in fin.readlines():
    obj=re.search("(X PIN)(\d*)(\s)(\d*)(\s.*)",ln)
if obj:
    num = int(obj.group(2))+25
    ln=obj.group(1) + str(num) + obj.group(3) + str(num) + obj.group(5) +'\n'
    fout.write(ln)
fin.close(); fout.close()
#
# for more info about regular expression syntax and KiCad component generation:
# http://gskinner.com/RegExr/
# http://kicad.rohrbacher.net/quicklib.php
```

1. While merging the two components into one, it is necessary to use the Library Editor from Eeschema to move the first component so that the second does not end up on top of it. Below you will find the final .lib file and its representation in *Eeschema*.

Contents of a *.lib file

```
EESchema-LIBRARY Version 2.3
#encoding utf-8
# COMP
DEF COMP U 0 40 Y Y 1 F N
F0 "U" -1800 -100 50 H V C CNN
F1 "COMP" -1800 100 50 H V C CNN
DRAW
S -2250 -800 -1350 800 0 0 0 N
S -450 -800 450 800 0 0 0 N
X PIN1 1 -2550 600 300 R 50 50 1 1 I

...

X PIN49 49 750 -500 300 L 50 50 1 1 I
ENDDRAW
ENDDEF
#End Library
```

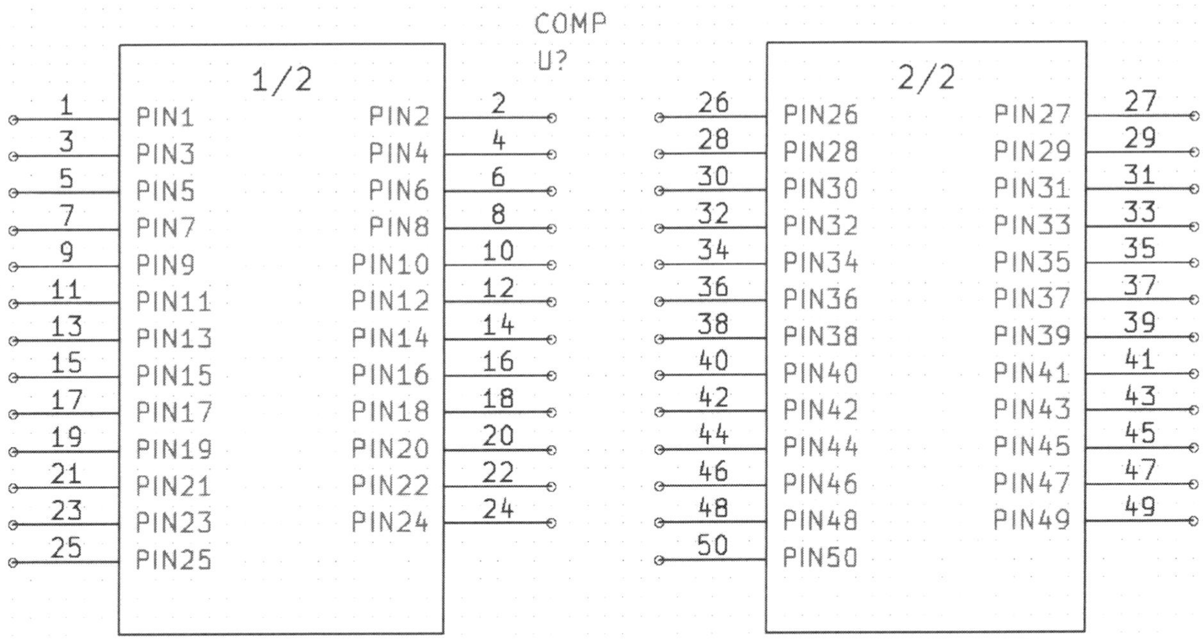

1. The Python script presented here is a very powerful tool for manipulating both pin numbers and pin labels. Mind, however, that all its power comes for the arcane and yet amazingly useful Regular Expression syntax: *http://gskinner.com/RegExr/*.

Chapter 7

Make component footprints

Unlike other EDA software tools, which have one type of library that contains both the schematic symbol and the footprint variations, KiCad *.lib* files contain schematic symbols and *.kicad_mod* files contain footprints. *Cvpcb* is used to successfully map footprints to symbols.

As for *.lib* files, *.kicad_mod* library files are text files that can contain anything from one to several parts.

There is an extensive footprint library with KiCad, however on occasion you might find that the footprint you need is not in the KiCad library. Here are the steps for creating a new PCB footprint in KiCad:

7.1 Using Footprint Editor

1. From the KiCad project manager start the *Pcbnew* tool. Click on the *Open Footprint Editor* icon on the top toolbar. This will open the *Footprint Editor.*
2. We are going to save the new footprint *MYCONN3* in the new footprint library *myfootprint.* Create a new folder *myfootprint.pretty* in the *tutorial1/* project folder. Click on the **Preferences** → **Footprint Libraries Manager** and press *Append Library* button. In the table, enter "myfootprint" as Nickname, enter "${KIPRJMOD}/ myfootprint.pretty" as Library Path and enter "KiCad" as Plugin Type. Press OK to close the PCB Library Tables window. Click on the *Select active library* icon on the top toolbar. Select the *myfootprint* library.
3. Click on the *New Footprint* icon on the top toolbar. Type *MYCONN3* as the *footprint name.* In the middle of the screen the *MYCONN3* label will appear. Under the label you can see the *REF* label. Right click on* MYCONN3 *and move it above* REF*. Right click on *REF__**, select *Edit Text* and rename it to *SMD.* Set the *Display* value to *Invisible.*
4. Select the *Add Pads* icon on the right toolbar. Click on the working sheet to place the pad. Right click

on the new pad and click *Edit Pad*. You can otherwise use the e key shortcut.

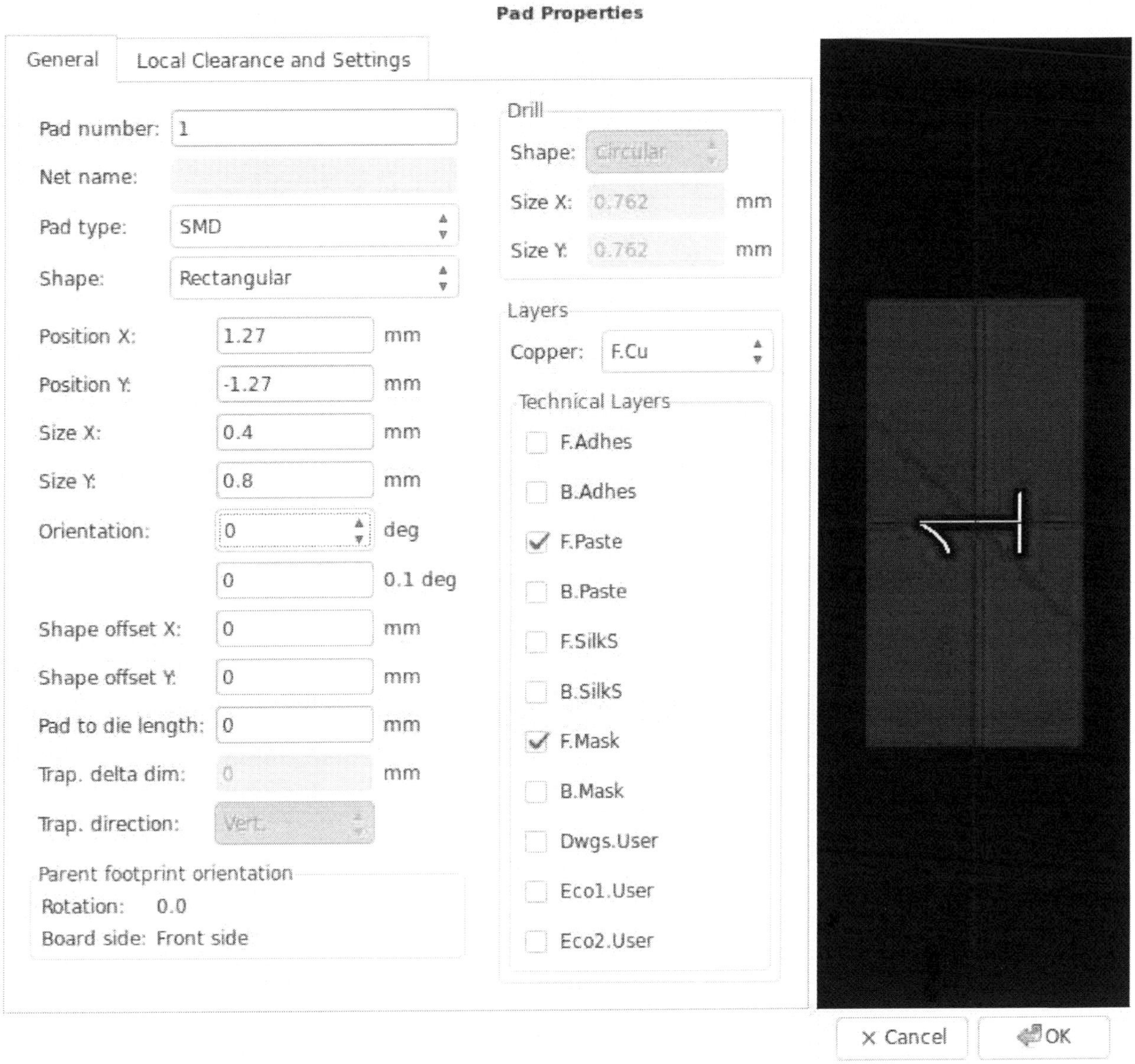

Pad Properties

General | Local Clearance and Settings

Pad number: 1
Net name:
Pad type: SMD
Shape: Rectangular
Position X: 1.27 mm
Position Y: -1.27 mm
Size X: 0.4 mm
Size Y: 0.8 mm
Orientation: 0 deg
0 0.1 deg
Shape offset X: 0 mm
Shape offset Y: 0 mm
Pad to die length: 0 mm
Trap. delta dim: 0 mm
Trap. direction: Vert.

Parent footprint orientation
Rotation: 0.0
Board side: Front side

Drill
Shape: Circular
Size X: 0.762 mm
Size Y: 0.762 mm

Layers
Copper: F.Cu

Technical Layers
F.Adhes
B.Adhes
F.Paste
B.Paste
F.SilkS
B.SilkS
F.Mask
B.Mask
Dwgs.User
Eco1.User
Eco2.User

× Cancel OK

5. Set the *Pad Num* to *1*, *Pad Shape* to *Rect*, *Pad Type* to *SMD*, *Shape Size X* to *0.4*, and *Shape Size Y* to *0.8*. Click OK. Click on *Add Pads* again and place two more pads.

6. If you want to change the grid size, **Right click** → **Grid Select**. Be sure to select the appropriate grid size before laying down the components.

7. Move the *MYCONN3* label and the *SMD* label out of the way so that it looks like the image shown above.

8. When placing pads it is often necessary to measure relative distances. Place the cursor where you want the relative coordinate point *(0,0)* to be and press the space bar. While moving the cursor around, you will see a relative indication of the position of the cursor at the bottom of the page. Press the space bar at any time to set the new origin.

9. Now add a footprint contour. Click on the *Add graphic line or polygon* button  in the right toolbar. Draw

an outline of the connector around the component.

10. Click on the *Save Footprint in Active Library* icon on the top toolbar, using the default name MYCONN3.

Chapter 8

Note about portability of KiCad project files

What files do you need to send to someone so that they can fully load and use your KiCad project?

When you have a KiCad project to share with somebody, it is important that the schematic file *.sch*, the board file *.kicad_pcb*, the project file *.pro* and the netlist file *.net*, are sent together with both the schematic parts file *.lib* and the footprints file *.kicad_mod*. Only this way will people have total freedom to modify the schematic and the board.

With KiCad schematics, people need the *.lib* files that contain the symbols. Those library files need to be loaded in the *Eeschema* preferences. On the other hand, with boards (*.kicad_pcb* files), footprints can be stored inside the *.kicad_pcb* file. You can send someone a *.kicad_pcb* file and nothing else, and they would still be able to look at and edit the board. However, when they want to load components from a netlist, the footprint libraries (*.kicad_mod* files) need to be present and loaded in the *Pcbnew* preferences just as for schematics. Also, it is necessary to load the *.kicad_mod* files in the preferences of *Pcbnew* in order for those footprints to show up in *Cvpcb*.

If someone sends you a *.kicad_pcb* file with footprints you would like to use in another board, you can open the Footprint Editor, load a footprint from the current board, and save or export it into another footprint library. You can also export all the footprints from a *.kicad_pcb* file at once via **Pcbnew** → **File** → **Archive** → **Footprints** → **Create footprint archive**, which will create a new *.kicad_mod* file with all the board' s footprints.

Bottom line, if the PCB is the only thing you want to distribute, then the board file *.kicad_pcb* is enough. However, if you want to give people the full ability to use and modify your schematic, its components and the PCB, it is highly recommended that you zip and send the following project directory:

```
tutorial1/
|-- tutorial1.pro
|-- tutorial1.sch
|-- tutorial1.kicad_pcb
|-- tutorial1.net
|-- library/
|   |-- myLib.lib
|   |-- myOwnLib.lib
|   \-- myQuickLib.lib
|
|-- myfootprint.pretty/
|   \-- MYCONN3.kicad_mod
```

```
|
\-- gerber/
    |-- ...
    \-- ...
```

Chapter 9

More about KiCad documentation

This has been a quick guide on most of the features in KiCad. For more detailed instructions consult the help files which you can access through each KiCad module. Click on **Help** → **Manual**.

KiCad comes with a pretty good set of multi-language manuals for all its four software components.

The English version of all KiCad manuals are distributed with KiCad.

In addition to its manuals, KiCad is distributed with this tutorial, which has been translated into other languages. All the different versions of this tutorial are distributed free of charge with all recent versions of KiCad. This tutorial as well as the manuals should be packaged with your version of KiCad on your given platform.

For example, on Linux the typical locations are in the following directories, depending on your exact distribution:

```
/usr/share/doc/kicad/help/en/
/usr/local/share/doc/kicad/help/en
```

On Windows it is in:

```
<installation directory>/share/doc/kicad/help/en
```

On OS X:

```
/Library/Application Support/kicad/help/en
```

9.1 KiCad documentation on the Web

Latest KiCad documentations are available in multiple languages on the Web.

http://kicad-pcb.org/help/documentation/

Made in the USA
San Bernardino, CA
28 February 2017